MICROBIOLOGY LABORATORY MANUAL

Second Edition
Revised Printing

GAYNE BABLANIAN
JEANIE PAYNE
Bergen Community College

Kendall Hunt
publishing company

Kendall Hunt
publishing company

www.kendallhunt.com
Send all inquiries to:
4050 Westmark Drive
Dubuque, IA 52004-1840

Copyright © 2003, 2008 by Dr. Gayne Bablanian & Dr. Jeanie Payne
Second Printing 2016

ISBN 978-1-5249-0358-9

Printed in the United States of America

CONTENTS

PART IV

MICROORGANISMS AND DISEASE   121

PART V

MICROORGANISMS IN THE ENVIRONMENT   159

PART VI

IDENTIFICATION OF UNKNOWN BACTERIA   197

PART VII

CONTROL OF MICROBIAL GROWTH   231

APPENDICES   291

LABORATORY ORIENTATION AND BASIC TECHNIQUES

Exercise 1   GENERAL RULES OF THE LAB

I. LABORATORY RULES AND SAFETY PRECAUTIONS

Laboratory work can be exciting and full of adventures, but it can also be dangerous if the proper precautions are not strictly followed. The following rules and regulations are important because they will ensure your safety.

1. Before you begin any experiment make sure you know the location of the fire alarms/extinguishers, fire blankets, eyewash stations, telephones and exits from the laboratory and the building.

2. It is suggested that students wear safety glasses when lighting the Bunsen burners. If you wear glasses made of safety glass, they are acceptable. Safety glasses may be purchased at any hardware store or in the bookstore.

3. Surgical gloves are recommended for most biology classes. They may be purchased at most drug stores.

4. Disinfect your work table before and after each lab.

5. Wash your hands with the hand cleaner provided before and after each laboratory session and before taking a break.

6. No eating or drinking in the laboratory.

7. Never bring unnecessary items into the laboratory. All coats must be hung on the coat rack. All books, pocketbooks, etc. must be in the desk slots at each lab station. Items too big to be placed in these slots, must be pushed under the lab desk and touch the center post. Nothing should be in your desk that is not needed for the lab experiment.

8. Always arrive to lab on time. Instructions are given at the beginning of the lab session. If you are habitually late, you may be asked to drop the course. It is dangerous to work in lab when you have not heard the instructions.

9. Read the lab experiment prior to lab time. Many mistakes are made when you don't know what you are supposed to be doing. Listen carefully to the instructor for modifications that are to be made.

10. Carry all equipment properly. Never try to carry two microscopes at a time, or your microscope and your books at the same time.

11. When you have finished an experiment, return all lab equipment to its proper place. Dispose of contaminated equipment to its proper place. Dispose of contaminated equipment in the appropriate receptacles. Your instructor will notify you of the glassware that may be safely washed and returned to the cabinets. In case of doubt, ask questions.

12. Never put anything in your mouth in the laboratory room. Pi pumps are used when pipetting is done. When inserting a pipette into a Pi pump use caution so as not to break the pipette. Excessive force is not needed, gently push the pipette into the Pi pump to obtain a seal.

13. Permanent marking pens or china markers are used to label test tubes, beakers, etc., and not gummed labels or scotch tape. It is recommended that you purchase a permanent marking pen.

14. When Bunsen burners, or open fires of any kind are used, individuals with long hair must tie back their hair.

15. No loose, dangling jewelry should be worn to lab. Chains etc. should be tucked into your clothing.

16. No floppy sleeves or other clothing that could get caught in the equipment should be worn to the lab.

17. No open toed shoes, sandals or opened backed shoes are allowed in the laboratory. No heels over 2 inches are allowed.

18. It is highly recommended that old clothes be worn to the laboratory. The school is not responsible if your clothes get ruined by dyes or chemicals used in the lab. It is recommended that you purchase a lab coat to protect your clothing. You may also purchase scrubs if desired.

19. Make sure that you read all labels carefully before using a chemical. Never put a chemical back into a stock bottle. Dispose of excess chemicals as instructed by your laboratory instructor.

20. **Report all accidents immediately to your instructor.** Remember it is critical for you to think before you do anything in a lab situation. Ask questions if in doubt. Accidents happen when carelessness or sloppiness occurs.

II. DRAW A MAP OF THE LABORATORY

To familiarize yourself with the lab room, find the items listed below. Label them on the blank diagram of the laboratory in Fig. 1–1.

Fire Extinguisher
Eye Wash Station
Broken Glass Disposal Container
Biohazard Disposal Container
Trash Barrel
Beakers and Test Tubes
Microscopes
Lens Cleaner and Lens Paper
Coat Rack
Exits
Hand Washing Sinks
Hot Plates
Experiment Set Ups
Glass and Hand Soap
Test Tube Brushes
Stains
Staining Sinks
Clean Glass Slides
Cover Slips
Oil (for oil immersion)
Bunsen Burners
Flint Lighters
Propipettes
Disposable Pipettes
Water Baths
Sterile Petri Dishes
Loops

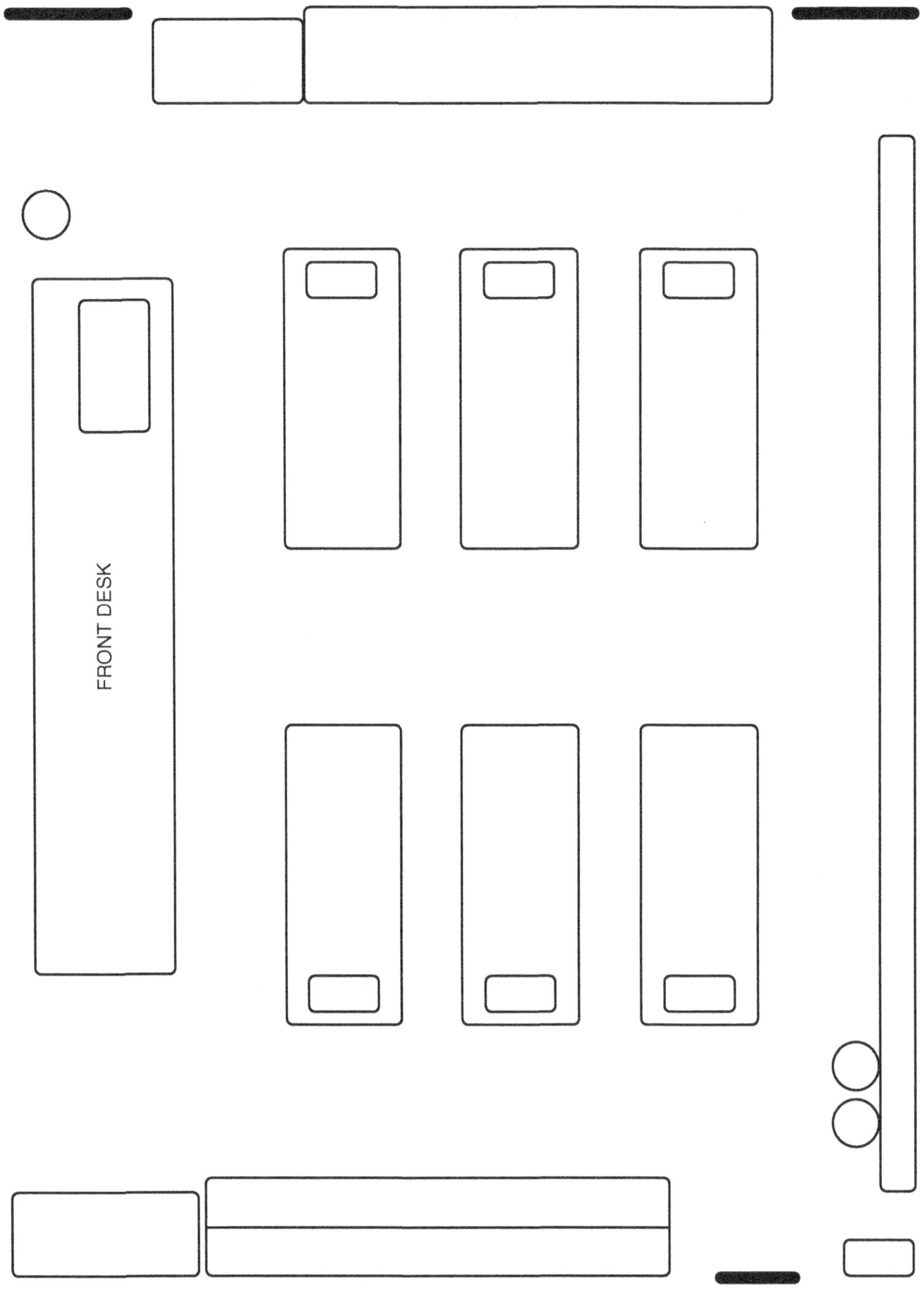

FRONT DESK

FIGURE 1–1

NOTES

Exercise 2   THE COMPOUND LIGHT MICROSCOPE

I. INTRODUCTION

The microscope is an invaluable tool used for observing objects that can not be seen by the naked eye. It enables us to view the **ultrastructure** (or fine detail) of organisms, tissues, and cells. In this lab exercise, you will be introduced to two types of microscopes.

The **compound light microscope** can magnify objects from 40 to 1000 times their original size. The specimens observed must be either small objects or very thinly cut sections of an object. In either case, these specimens must be translucent, allowing light to be transmitted through them.

The **dissecting (stereoscopic) microscope** magnifies objects in a range from 4 to 50 times. The specimens are usually opaque or too large to be viewed with the compound light microscope.

II. THE COMPOUND LIGHT MICROSCOPE

The compound light microscope utilizes an illuminating system and an imaging system which enable us to examine biological specimens. The illuminating system consists of a **light source (or illuminator)** that provides the light necessary to view the object. The **condenser** is a lens, located below the specimen, which condenses or focuses the light into a single, strong beam. Below the condenser is the **iris diaphragm,** which regulates the amount of light reaching the specimen.

The imaging system consists of a series of lenses and a body tube. The **objective lenses** are mounted on a revolving nosepiece. The magnifying power of each lens is 4X, 10X, 40X, and 100X. Notice that the magnification value is stamped on the side of each lens. Also, notice that each lens has a different length. The 4X objective is the shortest. The 100X lens is the longest. The higher the magnification of the lens, the shorter is the **working distance** (the distance between the lens and the specimen). The objective lenses of each microscope are also **parfocal,** which means when an object is in focus with one objective it is also approximately in focus with any other objective lens. The **ocular,** or eyepiece, has a magnification of 10X. Some microscopes are binocular (having two oculars).

The combined effect of the illuminating and imaging systems is the detection of very small objects. Image quality is dependent on a property of all microscopes called **resolution** or **resolving power.** Resolution is the ability of a microscope to distinguish fine detail in a specimen. The magnification of an object, and the amount of light available, enables us to see fine detail, or to clearly distinguish two points that are close together.

III. IDENTIFYING PARTS OF THE COMPOUND LIGHT MICROSCOPE

Locate and **identify** the following parts of the microscope in **Figure 2–1.**

1. **Light Switch**–Located in the base of the microscope.
2. **Light Source** (Illuminator)–Provides the light that illuminates the specimen. This is located in the base of the microscope.
3. **Iris Diaphragm**–Enables the viewer to adjust the amount of light that reaches the specimen. This is located below the condenser. A short black lever opens or closes the diaphragm.
4. **Condenser**–Focuses or condenses the light. This is located underneath the hole in the stage.
5. **Condenser Control Knob**–Adjusts the height of the condenser. Use the metal knob, located in front of the course and fine adjustment knobs.
6. **Mechanical Stage**–A platform with a slide holder. Serves to hold slides in place.
7. **Stage Control Knobs**–Used to move the slide around on the stage. Located hanging underneath the stage. Note: there are two knobs that control movement in two directions.
8. **Objective Lenses**–Set of three or four lenses located directly above the stage. The magnification of each lens is engraved in the metal lens housing.
   a. **Scanning Power**–This lens magnifies 4X and is the shortest of the three lenses, for initial focusing and viewing.
   b. **Low Power**–This lens magnifies 10X.
   c. **High Dry Power**–This lens magnifies 40X.
   d. **Oil Immersion Lens**–This lens magnifies 100X.
9. **Revolving Nosepiece**–Allows the objectives lenses to move into position above the specimen. Connects the lenses to the lower end of the body of the microscope.
10. **Body Tube**–A metal casing through which light passes to the oculars.
11. **Ocular (Eyepiece)**–This lens has a magnification of 10X, and is located in the uppermost part of the microscope. One of the oculars has a pointer that can be moved by turning the tube the ocular is housed within. If you have a binocular microscope, you can adjust the distance between the lenses to accommodate your eyes.
12. **Arm**–The upright structure attached to the base used in carrying the microscope.
13. **Base**–The heavy, flat support on which the microscope rests.
14. **Coarse Adjustment Knobs**–Used to rapidly alter the distance between the objective lenses and the stage to focus on an object. These are the large knobs that extend from each side on the lower part of the arm.
15. **Fine Adjustment Knobs**–Used to slowly alter the distance between the objective lenses and the stage when using the longer objective lenses (40X and 100X). These are the smaller pair of knobs that extend from each side of the lower arm.

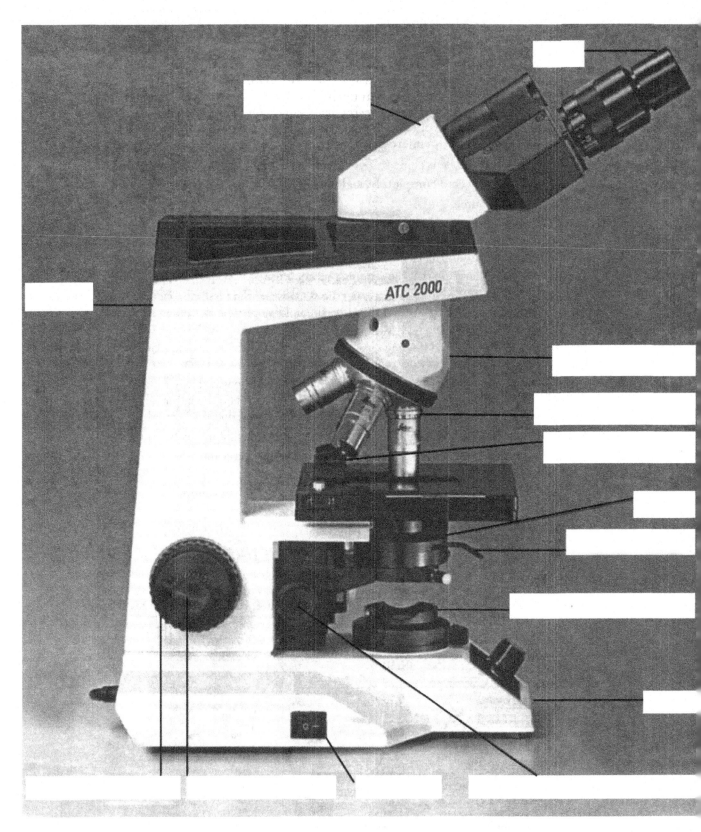

FIGURE 2–1

Photo courtesy of Leica Microsystems, Inc.

IV. USE AND CARE OF THE COMPOUND LIGHT MICROSCOPE

1. Carry the microscope to the lab bench using **TWO HANDS.** Grasp the arm of the microscope with your strongest (dominant) hand. Support the base with the other hand. Walk slowly and carefully to your table. Place the microscope down carefully. **[When you carry a microscope, this is the only object that should be in your hands. Do not attempt to carry slides or books along with microscope. Never tilt the microscope, as this may cause the ocular lens to fall out.]**

2. Clean all the lenses with lens paper.

3. Unwind the electrical cord **completely,** and plug it into an electrical outlet.

4. Turn on the light source.

5. With the condenser control knob, move the condenser up to the hole in the stage. You should see the light shining though the condenser. Open and close the iris diaphragm using the protruding lever. Note the change in light intensity. Now open the iris diaphragm completely to allow for a maximum light.

6. Turn the objective lenses so that the scanning power lens (4X) is clicked into place.

7. Obtain a prepared slide of the letter "e". On the stage is the slide holder. Open the metal arm to make room for your slide. Gently place the slide into the holder, nestling the slide towards the back. Gently allow the metal arm to close on the slide.

8. Practice moving the slide on the stage using the stage control knobs. Manipulate the slide so that the letter "e" is positioned directly over the condenser.

9. With the coarse adjustment knob, move the 4X objective all the way down, towards the slide. Now look through the ocular, and with the coarse adjustment knob, focus on the letter "e". (This is a good time to adjust the oculars for your eyes, if you have a binocular microscope.)

10. Move the "e" so that it is in the center of the field. Adjust the light intensity, if necessary.

   a. Move the slide to the LEFT. In which direction does the letter move?

   _____

   b. Move the slide to the RIGHT. In which direction does the letter move?

   _____

   c. Move the slide AWAY from you. In which direction does the letter move?

   _____

   d. Compare the orientation of the letter "e" when viewing it with the naked eye, then through the microscope.

   _____

   _____

   e. Draw the letter "e" below.

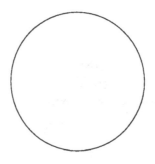

   Letter "e" Magnification _____ X

11. **WITHOUT TOUCHING THE COARSE FOCUSING KNOB,** turn the low power objective until it clicks into place. Using the coarse or fine adjustment knob, refocus on the letter. (Remember, the lenses are parfocal.)

    Draw the letter "e" below.

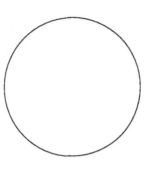

    Letter "e": Magnification _____ X

12. Before going to the next lens, position the "e" so that part of the letter is directly in the center of your field of view. Now turn the high power (40X) objective until it clicks into place. Using the **FINE ADJUSTMENT KNOB ONLY,** refocus on the letter.

    a. Draw the letter "e" below.

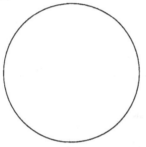

    Letter "e": Magnification _____ X

    b. What has happened to the field of view? _____

    c. What has happened to the letter "e"? _____

    d. What has happened to the light intensity? _____

    e. Why do you only use fine adjustment when viewing an object under high power?

    _____

13. The oil immersion lens is used for objects that need to be magnified 1000 times, such as bacteria. As its name implies, you need to use *immersion oil* with this lens. Immersion oil is especially formulated to have the same index of refraction as the glass from which the lens is made. Using the oil lowers the amount of light that would be lost otherwise due to the closeness of the lens to the slide.

14. Obtain a prepared slide of bacteria.

15. Repeat steps 6–12, focusing on the bacterial cells using the 4X, 10X and 40X objective lenses, in that order.

16. Before going to the 100X lens, position some of the bacterial cells directly in the center of your field of view.

17. Now move **HALFWAY** between the 40X objective lens and the 100X objective lens. This will leave a space over your slide. This gives you room to apply the immersion oil onto the slide.

18. Generally, you want to use a medium drop of oil. Let the excess oil drip off the applicator, and then touch the applicator to the slide.

19. Now turn the oil immersion lens (100X) until it clicks into place. Using the **Fine Adjustment Knob Only,** refocus on the bacterial cells.

20. Draw the cells below.

    bacillus   cocci   spirals

21. When you finished using the microscope for the day, complete the following steps:

    a. **Swing the oil immersion lens to the side and wipe the oil off using lens tissue *only*.**

    b. Turn the scanning power lens into place and raise it up.

    c. Return all slides to the proper slide trays.

    d. Turn off the light and unplug the cord.

    e. Clean the objective and ocular lenses.

    f. Drop the revolving nosepiece down toward the stage.

    g. Wrap the cord around the base of the microscope.

    h. Place the microscope back into the cabinet.

From *Biological Investigations* by Gayne Bablanian. Copyright © 2002 by Gayne Bablanian. Reprinted by permission of Kendall/Hunt Publishing Company.

V. TOTAL MAGNIFICATION

The **total magnification** of the object can be calculated by multiplying the power of the ocular by the power of the objective.

Ocular	×		Objective	=	Total Magnification
_____		(Scanning)	_____		_____
_____		(Low)	_____		_____
_____		(High)	_____		_____
_____		(Oil)	_____		_____

VI. DEPTH OF FIELD

Some objects on a slide can be thick and have many layers. The **Depth of Field** of a lens is its ability to distinguish between and focus sharply on each layer. There is a relationship between depth of field and magnification.

1. Obtain a slide containing three colored threads that are mounted on top of each other. Using the procedure in Part V, focus on the threads using the scanning power lens. Are all three threads in focus at the same time?

    _____

2. Click into the low power lens, and focus up and down. Can you determine which thread is on top?

    _____

3. Click into the high power lens. Can all three threads be brought into focus at the same time?

    _____

From *Experiencing Biology: A Laboratory Manual for Introductory Biology*, 5th Edition. Copyright © 2000 by GRCC Biology, 101 Staff. Reprinted by permission of Kendall/Hunt Publishing Company.

4. Which of the three lenses provides the greatest depth of field?

_____

VII. FIELD AREA

1. Obtain a clear plastic ruler with a metric scale.
2. Place the ruler under the microscope. Using the low power (4X) objective, focus on the ruler.
3. The space between any two lines is 1 millimeter (mm). What is the diameter of the field under:
   a. Scanning Power 4X? _____
   b. Low Power 10X? _____
   c. High Power 40X? _____
4. What happens to the diameter of the field as the magnification increases?

_____

_____

VIII. PREPARATION OF A WET MOUNT SLIDE

An alternative to using a prepared slide is to make your own slide, if a biological specimen is available. The **wet mount slide** is a temporary slide and can be used on living or preserved materials.

1. Obtain a clean glass slide.
2. Place your specimen (a small amount is generally the rule) onto the center of the **slide.** (see Figure 2–2)

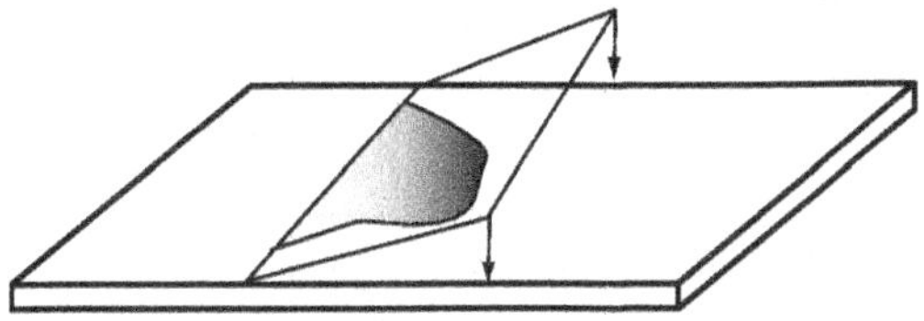

FIGURE 2–2

_____

3. Place the edge of a clean **coverslip** at the edge of the specimen. Slowly lower the coverslip over the specimen. This way, no air bubbles will get trapped.
4. When you are done viewing and drawing the specimen, discard the slide as instructed.
5. What is the purpose of using a coverslip?

_____

_____

_____

6. Make a wet mount slide and draw each of the following:
   a. *Elodea leaf:* Magnification _____ X

   b. Yeast: Magnification _____ X

   c. Cornstarch (add a drop of water to the slide): Magnification _____ X

IX. QUESTIONS

1. _____ The lens into which the observer looks is called the (1).
2. _____ The lenses on the revolving nosepiece are called the (2).
3. _____ Which objective should be in place when you first begin to focus?
4. _____ On low or medium power, which adjustment knob should be used for initial focusing?
5. _____ Which knob is used to make fine details sharply visible?
6. _____ Which is the only adjustment knob you use when looking through the high power objective?
7. _____ In which direction should you always focus, up or down?
8. _____ With what should you always clean a lens?
9. _____ The part of the microscope that controls the amount of light is called the (9).
10. _____ Which microscope provides a three-dimensional image?
11. _____ What property of a microscope allows it to remain in focus when switching objectives?
12. _____ If you are using a microscope having a 5X ocular and a 10X objective, what is the total magnification?
13. _____ When you are finished with the compound microscope, which lens should be left in position?
14. Describe the function of each of the following parts of the microscope: objective lens, ocular lens, coarse adjustment, fine adjustment, iris diaphragm.

_____
_____
_____
_____
_____
_____
_____
_____
_____

15. Explain the difference between magnification and resolving power.

_____
_____
_____
_____
_____
_____
_____
_____
_____

NOTES

Exercise 3   MICROSCOPIC MEASUREMENT

I. INTRODUCTION

Because size can be used to help distinguish different species of microorganisms, it is one of the most important physical characteristics that can be observed by a microbiologist. Since most organisms that you will be studying this quarter are microscopic, a specific piece of equipment known as an **ocular micrometer** can be used for measurement. The micrometer is a circular disk of glass, with graduations engraved on the surface, which has been placed in the ocular tube of the microscope (See **Figure 3–1**).

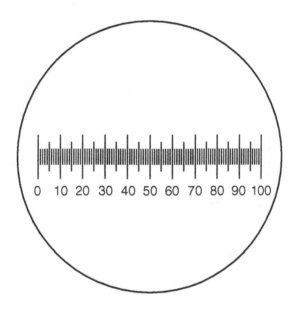

FIGURE 3–1
Ocular Micrometers

In order to use the ocular micrometer the distance between each graduation on each objective must be known. That is, it must be calibrated. The ocular micrometer can be calibrated using a stage micrometer which has the graduation distances measured precisely. To save time, this calibration procedure has already been completed and the calibrated distances for the ocular micrometer are given in **Table 3–1.**

From *Microbiology: Laboratory Manual* by Jake H. Barnes and Randall M. Brand. Copyright © 1995 by Jake H. Barnes and Randall M. Brand. Reprinted by permission of Kendall/Hunt Publishing Company

TABLE 3–1
Ocular Micrometer Graduation Distances

Magnification	Ocular Micrometer
Low Power (100X)	One graduation = 10 μm
High Dry (400X)	One graduation = 2.5 μm
Oil immersion (1000X)	One graduation = 1.0 μm

As an example, let us measure the length of a ***Paramecium.*** The slide containing the organism is first placed on the stage of the microscope and observed on low power. See **Figure 3–2.**

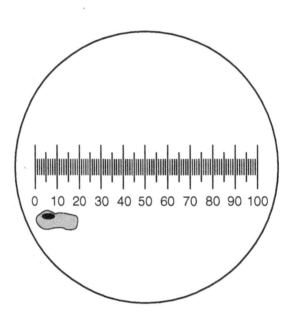

FIGURE 3–2
Paramecium Measurement

Notice that the ***Paramecium*** is about 15 graduations long on the ocular micrometer scale. If we know that one ocular micrometer graduation is equal to 10 μm (See **Table 3–1**), then all we have to do to determine the length of ***Paramecium*** is multiply the number of graduations (15) occupied by the specimen times the graduation distance (10 μm for low power) for the objective being used. So therefore, the ***Paramecium*** length is 15×10 μm $= 150$ μm.

If you wish to measure an organism on either High Dry or Oil Immersion, the same procedure must be followed. Remember to use the correct graduation distance for the objective that is used for measurement. Today's lab will give you practice in the use of the ocular micrometer for measurement.

From *Microbiology: Laboratory Manual* by Jake H. Barnes and Randall M. Brand. Copyright © 1995 by Jake H. Barnes and Randall M. Brand. Reprinted by permission of Kendall/Hunt Publishing Company.

II. MATERIALS

Microscope with Ocular Micrometer
Euglena Slide

III. PROCEDURE

1. Obtain one slide of *Euglena* from the preparation table.
2. Place the slide on the stage of your microscope and focus on low power. By using the mechanical stage of your microscope, align the *Euglena* specimen you wish to measure with the 0 on the ocular micrometer scale. (Note: the ocular can be rotated to align the scale with the specimen). Calculate the *Euglena* length by using the ocular micrometer graduation distance in **Table 3–2 Record your data in the table on the worksheet.**
3. Before you rotate your nosepiece to high dry, be sure to center the specimen around the tip of the pointer so that you know the exact location of the specimen to be measured. Now measure on high dry the length of the same *Euglena* that you measured on low power. **Record that information in the table on the worksheet.**
4. Repeat step 3 for oil immersion, then answer the questions on the data sheet.
5. Place the *Euglena* slide in the box on the preparation table. Clean your microscope and return it to the proper storage area.

From *Microbiology: Laboratory Manual* by Jake H. Barnes and Randall M. Brand. Copyright © 1995 by Jake H. Barnes and Randall M. Brand. Reprinted by permission of Kendall/Hunt Publishing Company.

Name _____   **Date** _____

TABLE 3–2

Magnification	# of graduations long	1 ocular graduation is equal to:	Euglena length (μm)
Low (10X)			
High dry (40X)			
Oil (100X)			

IV. QUESTIONS

1. Should the length of *Euglena* that you measured be consistent on all three objectives? Why?

2. If the organism shown below is measured on high dry using a different microscope and we know that 1 graduation on high dry is equal to 3 μm; then what is the length of the organism?

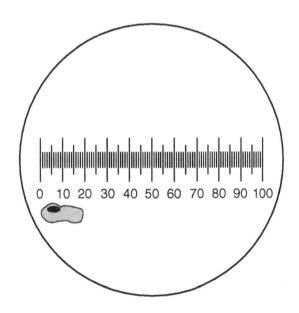

Exercise 4 CULTURAL CHARACTERISTICS OF BACTERIA

I. INTRODUCTION

Microorganisms (viruses, chlamydia, rickettsia, bacteria, cyanobacteria, fungi, molds, algae, and protozoa) are found everywhere in nature; from ocean depths to mountain peaks, from swamps to deserts, and from frigid polar regions to the super-heated water escaping from vents on the ocean floor. The environment where an organism is normally found is presumably the environment to which the organism is best adapted, i.e., that environment is the organism's ecological niche.

While individual microorganisms have particular ecological niches, they are not necessarily confined to those niches. In fact, a microorganism transferred to a new niche by accident or deliberate transfer may even find the new environment more adventitious than the environment that is its normal niche. The ability to tolerate, or even to take advantage, of new environments, coupled with the fact that microorganisms are very small and exist virtually everywhere, means that special procedures are necessary for handling them effectively. The nature of environmental influences and the background and detail of how microorganisms respond to environments is a major focus of the laboratory portion of this course. While more detail and discussion of why organisms exist where they do, this lab exercise can begin to demonstrate just how ubiquitous microorganisms really are.

II. COLONY MORPHOLOGY

Procedure:

1. Groups of four (4) students should obtain five (5) nutrient agar plates (or 4 TSA plates and 1 BAP) from the supply provided and handle them as described below.

    a. Open one plate and expose it to the air for 30 minutes.

    b. Each person should touch a finger to a marked quadrant of one of the plates (everyone touch a different portion of the same plate!)

    c. One individual should cough or sneeze on a blood agar plate.

    d. Another individual should hold his/her head over a plate and vigorously comb his/her hair.

    e. Swab some surface (other than a laboratory desk top) with a sterile swab, then streak (rub) the swab onto the remaining agar plate.

    Incubate plates (lid down, or inverted) in your drawer until the next lab period.

2. After incubation, observe your plates for different colony types. Focus upon colony surface form and texture, elevation, and margin (see Fig. 4–1), and describe several colony types in Table 4–1 on the next page.

3. Practice making other discriminating observations like counting the number of colonies of each type on a given plate, and comparing colony types on the different plates.

4. Someone in the group should save one plate containing colonies so changes in colony appearance can be followed for several days. The other plates should be discarded as directed.

Taken from General Microbiology (3rd Edition), by James E. Urban, Pgs. 9–14.

From *Urban General Microbiology*, Third Edition by James E. Urban. Copyright © 1996 by James E. Urban. Reprinted by permission of Kendall/Hunt Publishing Company.

TABLE 4–1

Summary of Microorganism Omnipresence Data

Environment Sampled	Colony Characteristics	Colony Color	Number of Colonies

1. PLATES (Colonies)
   a) *General surface form*

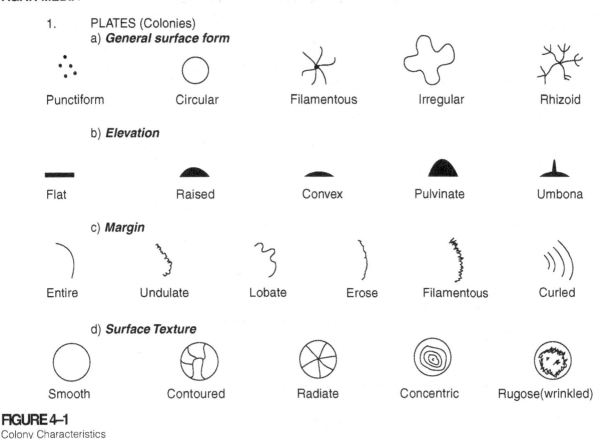

FIGURE 4–1
Colony Characteristics

III. CELL MORPHOLOGY AND CELL GROUPING

All bacterial cells have a characteristic and genetically determined cell shape. In the eubacteria (true bacteria,) cell shape and is a function of the chemical arrangement of the peptidoglycan (murein) which makes up the cell wall of Gram (+) bacteria or the Gram (−) bacteria. The overwhelming majority of bacteria and eubacteria and have either the bacillus, coccus, vibrio, spirillum, or spirochete shape (See Fig. 4–2). The bacillus (rod) and coccus (sphere) shapes are typically clear and easily recognized, but the spiral shapes [vibrio (banana), spirillum (rigid spiral), and spirochete (flexible spiral)] are often difficult to differentiate. In addition, growth or environmental conditions can cause organisms to assume a less-distinct, pleomorphic, shape. Within the past few years some archaebacteria isolated from comparatively hostile environments have been found to display starshaped, triangular, and square morphologies. Archaebacteria do not have peptidoglycan in their cell walls and this presumably accounts for the different morphologies observed in those bacterial types.

Dependent upon the division planes of individual organisms, cell surface properties, and to some extent the degree of cellular motility, cells displaying characteristic cell shapes also display characteristic cell groupings. These groupings are singles, pairs, tetrads, sarcina, and staph-like clusters. Groupings are division-related (result from orientation of successive division planes and the cohesiveness properties of the surfaces of newly divided cells,) and post-divisional groupings (related primarily to cohesiveness properties of the cell surface, may also occur).

Procedure:

1. Obtain prepared slides of several species of bacteria.
2. Examine each slide. Make careful drawings below of cell shapes and groupings, using Fig. 4–2 as a guide.

CHARACTERISTICS AND TERMINOLOGY

Cell Morphology

Shape of vegetative cells	Terminology	Description
	bacillary	rod or cylinder
	coccobacillary	rod with tapered ends; short
	coccus	spherical to avoid may be flat on one side
	vibrio	comma or banana
	spirillum	rigid spiral; long wavelength
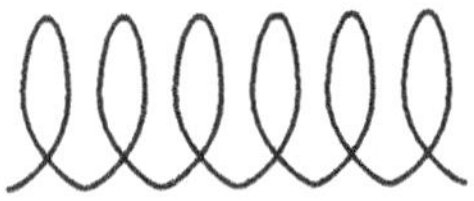	spirochete	flexible spiral, short wavelength; axistyle

Cell Grouping

Cell Grouping	Terminology	Description
	singles	no detectable grouping
	pairs	two attached cells
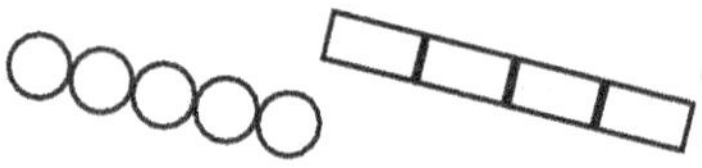	chains	more than two cells; one division plane
	tetrad	packages of four cells; two division planes
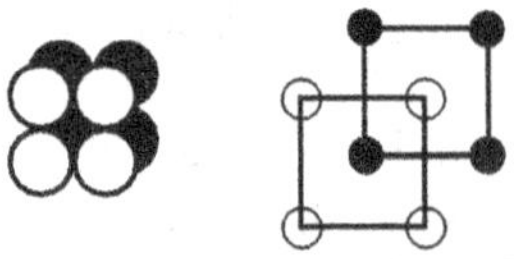	sarcina	packages of eight eight cells; three division planes
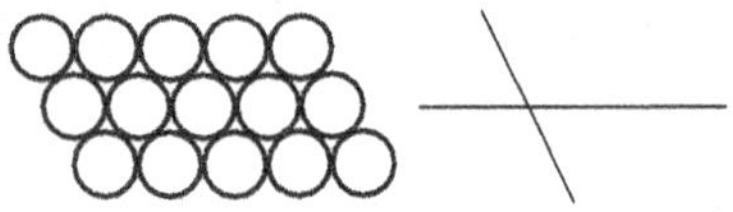	staph	irregular clusters; rows at acute angles

FIGURE 4–2

Characteristics and Terminology

From *Urban General Microbiology*, Third Edition by James E. Urban. Copyright © 1996 by James E. Urban. Reprinted by permission of Kendall/Hunt Publishing Company.

NOTES

Exercise 5   PREPARATION AND CARE OF STOCK CULTURES, AND ASEPTIC TECHNIQUE

I. INTRODUCTION

You will be using bacterial organisms for your unknown cultures and for studying various techniques in microbiology. Although you will practice the use of aseptic techniques, there is still a chance of contaminating the cultures when used frequently. For example: If you were given a tube of broth containing a known bacterial organism, and you were preparing all of your inoculations from that tube, eventually you would most likely contaminate the tube with another bacterial organism. In order to alleviate the chance of contamination you will practice aseptic techniques and you will use two different types of stock cultures.

There are two types of stock cultures: **(1) reserve stock culture and (2) working stock culture.** A **reserve stock culture** is stored in the refrigerator after incubation for later use. The reserve stock culture can be used to make transfers onto another reserve stock or to make a working stock culture. The **working stock culture** is used for preparing slides, making stains and for routine inoculations. If it becomes contaminated, you can replace it with a fresh culture from the reserve stock.

II. STERILE TRANSFER TECHNIQUE

Materials:
Disinfectant
Sponge
Test Tube Rack
Test Tubes W/Caps (2)
Agar slant
Test Tube with Nutrient Broth
Bunsen Burner

Cultures:
Serratia marcescens broth culture
Serratia marcescens slant

III. PROCEDURE

1. Using a bottle of disinfectant and a sponge provided on the preparation table, disinfect your work table. Return the bottle of disinfectant and sponge to the appropriate area.

2. As demonstrated by your instructor, exercise extreme caution in lighting your burner.

3. Obtain a test tube rack, two blank test tubes, and an inoculating loop from the preparation table.

4. Referring to the following illustrations in Fig. 5–1, practice your aseptic technique for bacterial transfer by using blank tubes for a "dry run". Have your lab partner observe your "dry run" until you have performed it correctly.

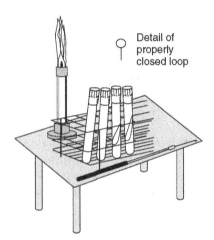

Detail of properly closed loop

a. Your rack and test tubes should be arranged as shown above.

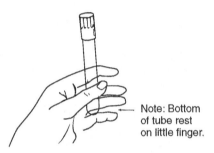

Note: Bottom of tube rest on little finger.

b. Hold both culture tube in your left hand. (Hold them in your right hand if you are left-handed.) The tubes should *not* be held vertically once the closures are removed.

c. Hold your inoculating loop in your right hand. (Hold loop in your left hand if you are left-handed.) The loop should be held like a pencil.

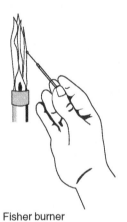

Fisher burner

d. Flame the inoculating loop in the Bunsen burner holding the loop parallel to the Bunsen burner so that all the wire gets red-hot at once. Allow the loop to cool so that you do not cremate the living bacterial cells you are about to transfer.

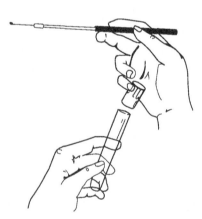

e. Remove the closure from culture tube. Do this by wrapping the little finger of your right hand around the closure of the tube nearest to your right hand. Grasp only the uppermost portion of the closure so that the open end does not touch the heel of your hand.

FIGURE 5–1

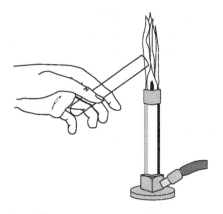

g. Flame the neck of the uncovered tube by passing it back and forth through the flame twice. Be careful to hold the tube in a nearly horizontal position.

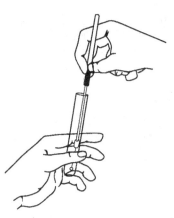

h. Insert the inoculating loop into the pure culture (or the practice tube substituting for it), and remove a small amount of bacteria. Note the position of the closures in the right hand. Reflame the neck of the pure culture tube, cap it and put it back in the rack.

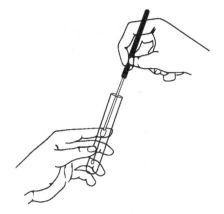

i. Pick up the tube of sterile media to be inoculated, flame the neck of that tube. Transfer this inoculum (the small amount of bacteria) to the surface of the uninoculated slant. (or the practice tube substituting for it).

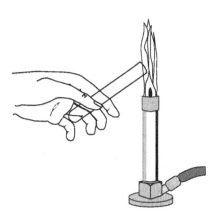

j. Reflame the neck of the tube.

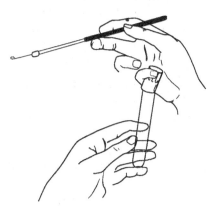

k. Recap the tube.

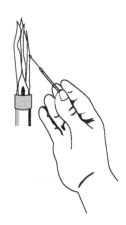

l. Reflame the loop, killing all the bacteria on it.

m. Return the tube and inoculating loop to the rack as shown in Part a.

From *Microbiology Laboratory Manual* by Jake H. Barnes and Randall M. Brand. Copyright © 1995 by Jake H. Barnes and Randall M. Brand. Reprinted by permission of Kendall/Hunt Publishing Company.

5. Each and every time you make an inoculation, you must label it carefully and completely. An adequate label should include your name, the date of inoculation, the name of the organism, the source of inoculum if appropriate (such as a throat swab), the type of medium, the temperature of incubation, and the exercise number. See Figure 5–2 for an example.

30 C	Jane Smith	9/12/90
	Serratia Marcescens	
Asepsis	Act. 2	NA

FIGURE 5–2

IV. TRANSFERRING CULTURES FROM A LIQUID CULTURE TO AN AGAR SLANT

1. You will need an unused agar slant for the transfer of the liquid culture.
2. Tap the tube containing the liquid culture so that the bacteria are evenly distributed in the culture.
3. Flame sterilize the inoculating loop. Do one tube at a time.
4. Pick up a broth tube with microbes and a new agar slant. Hold both tubes in the same hand. Remove the cap and avoid setting the cap down on the table. Flame the mouth of the tube and hold both tubes nearly horizontal to keep airborne microorganisms from entering.
5. Put the end of the loop into the broth tube that has bacteria growing in it. Remove the loop from the broth culture. You should see liquid on the loop.
6. Insert the loop with the microbes into a fresh agar slant. Gently streak your loop from the base of the agar slant to the top, in one continuous stroke. Do not gouge the agar. You might see some water at the bottom of the agar slant. This is condensation from when the agar cooled. Do not let the water run up over the surface of the agar slant when you do your streaking with the loop. (See Fig. 5–3)
7. Flame the mouth of the tube and immediately replace the caps. Put the tubes back in the test tube rack. Reflame the inoculating loop. Incubate at 25°C (on top of incubator).

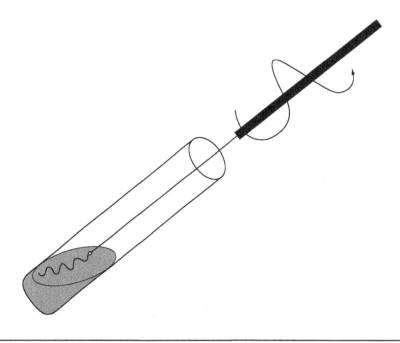

FIGURE 5–3

V. TRANSFERRING CULTURES FROM AN AGAR SLANT TO BROTH

1. You will need a sterile broth tube for transfer of the agar slant culture.
2. Flame sterilize a straight inoculating needle. Do one tube at a time.
3. Pick up and hold the tube with the microbes growing on agar and the new tube of broth in one hand. Your other hand will be holdsing the inoculating needle. Turn the agar slants so that the surface of the slants is facing the ceiling.
4. Remove the caps from the tubes with the hand in which you are holding the needle. Never put the caps on the table. Hold both tubes in the same hand. Flame the mouths of the tubes and hold both tubes nearly horizontal to keep airborne microorganisms from entering.
5. Gently touch the bacteria on the surface of the agar using the inoculating needle. Be sure not to gouge the agar.
6. Transfer the bacteria you have on the needle into the broth. Insert the needle a short distance into the liquid and scrape the end along the glass surface in the broth to release the bacteria.
7. Flame the mouth of the tube and immediately replace the caps. Put the tubes back in the test tube rack. Reflame the inoculating loop.
8. Tap the broth tube to distribute the microbes throughout the broth in the tube.
9. Turn off your burner and disinfect your work table. Place the stock culture in it's original container and your inoculated agar slant in the incubator. To obtain good growth, it should be incubated for 48 hours at 25 degrees C. Remember to wash your hands.

VI. LAB DAY TWO

1. Remove from the incubator your inoculated agar slant of *Serratia marcescens* and examine it for growth and possible contamination. Answer all questions in Section VII.
2. Discard your culture tube in the appropriate area as indicated by your instructor.

From *Biology 6 General Microbiology: Laboratory Manual* by Ken Kubo, Bob Lenn and Lori Smith. Copyright © 2001 by Ken Kubo, Lori A. Smith and Robin Lenn. Reprinted by permission of Kendall/Hunt Publishing Company.

VII.

Name _____   **Date** _____

1. Draw a diagram of your inoculated agar slant below.

2. Why is it important that a small amount of inoculum be used in preparation of an agar slant?

3. Why is it important that you disinfect your work area both before and after working with bacteria?

4. How could you determine if there is any contamination on your agar slant?

NOTES

STAINING AND SLIDE TECHNIQUES

Exercise 6   THE NEGATIVE STAIN

I. INTRODUCTION

In order to prepare a specimen for most staining protocols, the normal procedure is to allow the organisms to dry to a glass slide and then pass the slide through the Bunsen burner in order to fix the organisms to the slide. This procedure, however, does not allow one to visualize the cells in their natural state. The size and shape of the cell may become distorted during this procedure. Also, heat-fixing removes the water from the cell, thereby affecting the size of the cell when it is viewed under the microscope. Although these distortions may be acceptable under some circumstances, one may want to visualize microorganisms as they appear in their natural state.

Negative staining allows one to view living organisms that have not been heat-fixed to a slide. In addition, this type of staining allows one to visualize organisms or parts of organisms, such as a capsule, which may not readily take up stains. In this type of staining the background is filled with a stain such as India ink or nigrosin, while the organisms themselves are clear, unstained objects that stand out clearly against this dark background. These stains do not penetrate the bacterial cells due to the repulsion between the negative charge of the stains and the negatively charged bacterial cell.

In this exercise you will prepare negative stains utilizing two different methods. A suspension of bacterial cells is emulsified in a drop of nigrosion and covered with a cover slip (wet mount) or allowed to air dry (smear).

Materials:

Microscope slides

Nigrosine solution

Inoculating loops

Bunsen burner

China marking pencil or felt marking pen

Cultures:

Klebsiella pneumoniae

Bacillus megaterium

From *Exercises for the Microbiology Laboratory* by Elizabeth Fish McPherson. Copyright © 2001 by Elizabeth Fish McPherson. Reprinted by permission of Kendall/Hunt Publishing Company.

II. PROCEDURE

A. **Negative Stain—Wet Mount**: See Fig. 6–1

1. Wash and dry a microscope slide.

2. Place a drop of **nigrosin** in the center of this slide.

3. Aseptically transfer a sample of cells to the staining solution and mix gently. **See Fig. 6–2.** (Do not forget to flame the inoculating loop before and after use.)

4. Place a clean cover slip over the mixture of nigrosin and cells.

5. **Examine** the preparation microscopically using the **low power (10X)** and **high power (40X)** objectives **only.** The **cells** will appear **colorless (clear)** in a **gray** to **black background.** Draw the organisms on the Worksheet.

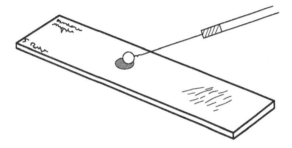

1. Wash and dry a microscope slide and place a drop of nigrosine in the center of the slide.

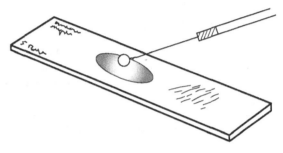

2. Transfer a sample of cells to the staining solution and mix gently.

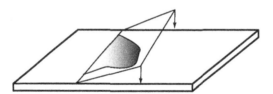

3. Place a clean cover slip over the mixture of nigrosine and cells and observe using (40X) and oil immersion (100X) objectives.

FIGURE 6-1

Material adapted from *Microbiological Laboratory Manual* by Brenda G. Wellmeyer, drawings adapted from *Microbiology Laboratory Manual* by Lois M. Clouser and *Biological Investigations* by Gayne Bablanian.

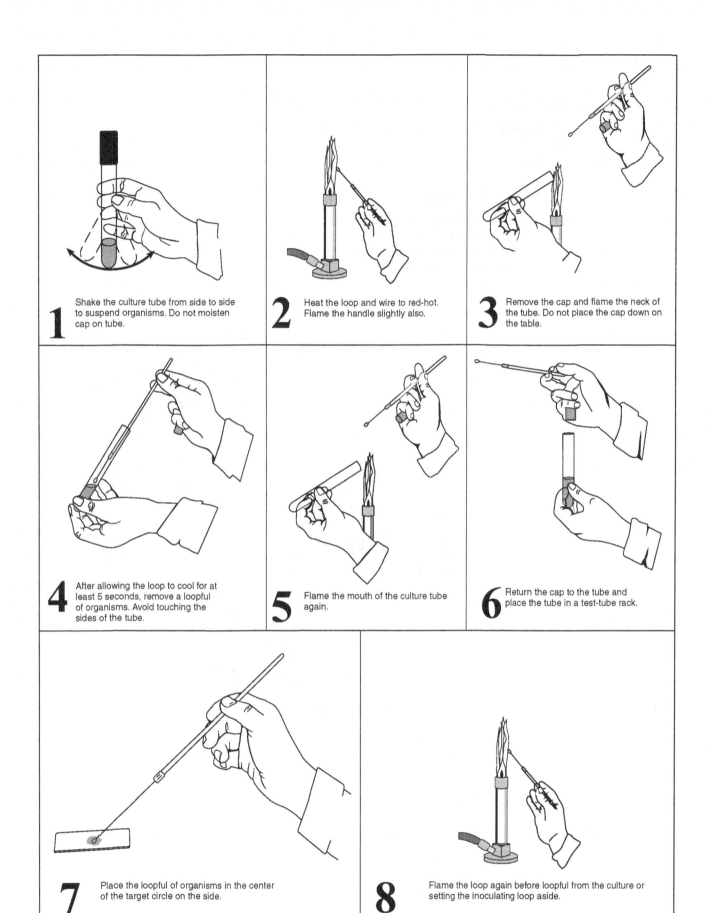

1 Shake the culture tube from side to side to suspend organisms. Do not moisten cap on tube.

2 Heat the loop and wire to red-hot. Flame the handle slightly also.

3 Remove the cap and flame the neck of the tube. Do not place the cap down on the table.

4 After allowing the loop to cool for at least 5 seconds, remove a loopful of organisms. Avoid touching the sides of the tube.

5 Flame the mouth of the culture tube again.

6 Return the cap to the tube and place the tube in a test-tube rack.

7 Place the loopful of organisms in the center of the target circle on the side.

8 Flame the loop again before loopful from the culture or setting the inoculating loop aside.

FIGURE 6–2

Material adapted from *Microbiological Laboratory Manual* by Brenda G. Wellmeyer, drawings adapted from *Microbiology Laboratory Manual* by Lois M. Clouser and *Biological Investigations* by Gayne Bablanian.

B. **Negative Stain—Smear:** See Figure 6–3.

1. Place a drop of **nigrosin** on a clean microscope slide.

2. Aseptically transfer a sample of cells to the staining solution. Gently mix the cells in the nigrosin **spreading** the stain and microorganisms over the surface of the slide as though you are preparing a smear.

3. Allow the smear to **air dry. Do not heat fix** the slide.

4. **Examine** the smear microscopically using the **low power (10X), high power (40X)** and **oil immersion (100X)** objectives. Place immersion oil on the surface of the dried smear. (Hint: If the dye is too thick it will crack as it dries. Observe the outer edges of the smear where the dye is thinner.) Draw the organisms on the Worksheet.

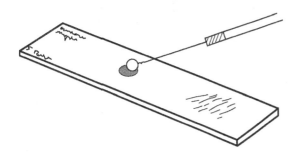

1. Place a drop of nigrosine on a clean microscope slide.

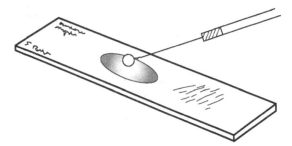

2. Transfer a sample of the cells to the staining solution and gently mix the cells in the nigrosine.

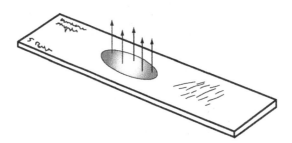

3. Spread the stain and the microorganisms over the surface of the slide as though you are preparing a smear. **Air Dry. Do not heat fix.** Examine under oil immersion (100X).

°(Hint: If the dye is too thick it will crack as it dries. Observe the outer edges of the smear where the dye is thinner.)

FIGURE 6–3

Material adapted from *Microbiological Laboratory Manual* by Brenda G. Wellmeyer, drawings adapted from *Microbiology Laboratory Manual* by Lois M. Clouser and *Biological Investigations* by Gayne Bablanian.

Name _____   **Date** _____

Examine the slides that you have prepared of the negative stain. Observe with high power (40X) and oil immersion (100X) objectives. Record your observations in the spaces below.

Negative Stain:
Wet Mount
(40×)

Negative Stain:
Smear
(100×)

QUESTIONS

1. Why is nigrosin used as a negative stain?

   _____

   _____

   _____

   _____

   _____

2. Why is the slide not heat fixed before negative staining?

   _____

   _____

   _____

   _____

Material adapted from *Microbiological Laboratory Manual* by Brenda G. Wellmeyer, drawings adapted from *Microbiology Laboratory Manual* by Lois M. Clouser and *Biological Investigations* by Gayne Bablanian.
Questions taken from Microbiology: Laboratory Manual, Second Edition, by Frank H. Osborne, Pg. 29.

Exercise 7   PREPARING A BACTERIAL SMEAR

I. INTRODUCTION

Most microorganisms appear colorless when viewed through a brightfield light microscope. Therefore, special steps must be taken in order to observe microorganisms under a microscope. The most common method of allowing visualization of microorganisms under a microscope is to **stain,** or color them. Staining imparts color to the microorganisms so that various structures can be observed.

One of the more important techniques you will learn in this laboratory is how to properly prepare a bacterial **smear** for viewing with the brightfield microscope. If a smear is prepared improperly, trying to view it under the microscope will be useless. Smears are often difficult for beginners to master. Practice is the key to preparing good smears.

When preparing a smear, a drop of liquid bacterial culture or a drop of water and a portion of a bacterial **colony** (a large number microorganisms growing as a solid mass on an agar medium) is spread over the surface of the slide. If too many bacteria are present in the smear, clear differentiation of individual cells when observed under the microscope will be difficult. On the other hand, if too few bacteria are present in the smear, the microscopist will have to spend a long time searching the slide for a bacteria cell to view. One loopful of liquid bacterial culture usually provides sufficient bacteria for easy viewing. When preparing a smear from a bacterial colony, it is important to remember that one bacterial colony contains millions of individual bacteria. A tiny portion of the colony will suffice. Additionally, when mixing water with the bacteria to make a smear, only a small drop of water is necessary. Because the smears are air-dried, the more liquid that is present, the longer the smear will take to dry. Finally, when mixing the bacteria with the water to create the smear, avoid overmixing because this can lead to inaccurate bacterial cell patterns (chains may appear as clumps and vice versa).

Once the smear has air-dried, the next step is to **heat-fix** the slide. In this procedure the air-dried smear is briefly passed through the Bunsen burner flame two or three times. This process serves to kill the bacteria in the smear, causes the bacteria to adhere to the slide, and causes changes in the bacteria cells which enable them to stain more readily. This step can also present some problems. The purpose of heat-fixation is to gently heat the slide to kill the cells, not to incinerate the cells. Therefore, holding the slide in the flame for a long period of time is not acceptable. Incinerated cells will not stain and nothing will be seen under the microscope. However, the smear must be heated enough so that the bacteria cells are killed. If the bacteria are not killed, their ability to accept stains is altered. Finally, trying to rush through the procedure and heat-fix a smear before it is completely air-dried results in the organisms being boiled and no cells will adhere to the slide. As was stated earlier, practice is of utmost importance in preparing smears.

Materials:

Microscope slides
Bunsen burner
Inoculating loop
China marking pencil
clothespin

Cultures:

Broth cultures of yeast (optional), *Bacillus megaterium, Staphylococcus aureus* or other broth samples
Nutrient agar slant cultures of *Bacillus megaterium, Staphylococcus aureus* or other slant cultures

II. PROCEDURES FOR SMEAR PREPARATION

A. From Solid Media (Figure 7–1):

The following procedure describes the preparation of a smear from a solid medium (slant cultures).

1. Thoroughly **wash** and dry the slides. Use warm water and the detergent provided. (**Do not use** the hand washing liquid.) Do not touch the slides with your fingers after they are clean.

2. **Label** your clean slides with the name of the microorganism. All slides should be labeled **before** you begin preparing the smears.

3. Using a flamed loop place a few drops on the slide.

4. Following the procedure from Figure 6–2, aseptically transfer a **small amount** of growth from the slant to the drop of water. Gently **emulsify** the microbes in the water. Put the slide on a dark surface, such as the lab table, to provide greater contrast.

5. **Spread** the emulsified microbes and water over an area on the slide approximately the **size of a nickel** in order to produce a **monolayer of cells.** If the smear appears milky or white against the dark tabletop, you are using too many microorganisms. **Flame** the inoculating loop before putting it down.

6. Allow the smear to **air dry** until **all the water** evaporates. Placing the slide in a warm spot near the Bunsen burner may speed up this process. When dry, the smear should appear to be **very thin** or barely visible.

7. **Heat fix** the slide by passing the slide **film side up** across the top of the Bunsen burner flame two or three times. You may grasp the end of the slide with a slide holder (or clothespin) to protect your fingers from the heat. **Do not over heat the slide.** Excess heat will cause lysis of the cell walls. The mild heat will coagulate cell proteins causing the bacteria to **adhere** to the slide and resist being washed off during the staining process.

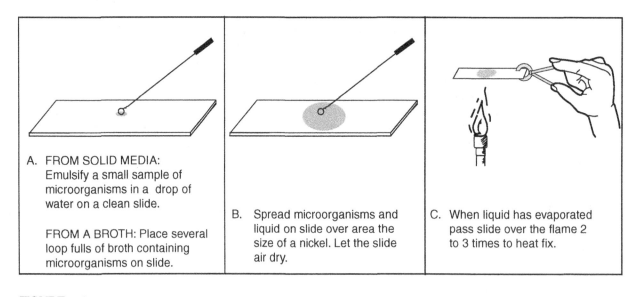

A. FROM SOLID MEDIA: Emulsify a small sample of microorganisms in a drop of water on a clean slide.

FROM A BROTH: Place several loop fulls of broth containing microorganisms on slide.

B. Spread microorganisms and liquid on slide over area the size of a nickel. Let the slide air dry.

C. When liquid has evaporated pass slide over the flame 2 to 3 times to heat fix.

FIGURE 7–1
Preparation of a Smear

B. **From Broth**:

The following describes the preparation of a smear from a broth culture.

1. Aseptically remove several loopfuls of microorganism suspended in a broth culture.
2. Spread the microorganisms and broth directly on a clean slide.
3. The slide should be air-dried and heat fixed as previously described.

QUESTIONS

1. Why is it necessary to heat the slide after the smear is air-dried? Give two reasons.

_____

_____

_____

_____

_____

_____

2. How much bacteria should be spread over the surface of the slide when preparing a smear?

_____

_____

_____

_____

_____

_____

NOTES

Exercise 8   THE SIMPLE STAIN

I. INTRODUCTION

Staining a smear of bacteria simply adds color to bacteria to make them easier to see using a microscope. The stain is a dye which binds to a cellular structure and allows the bacteria to appear as colored objects against a clear background. There are two different types of stains used in microbiology. **Simple stains** consist of a single dye which is added to a smear.

The procedure involves the use of a basic dye that is positively charged. Bacterial cell walls are negatively charged and are readily stained by the basic dyes (stains). This type of stain is used when information about cell shape and size, as well as the arrangement of the bacterial cells, is wanted. All bacteria take up the stain, so one cannot discern differences in cell wall structures. Common simple stains are methylene blue, safranin, carbol-fuchsin, crystal violet, and malachite green.

Note: In contrast **differential stains** are composed of two or more dyes and can differentiate between organisms and cell wall structures. This type of staining highlights the differences between bacteria and can be used as a classification tool. The most common type of differential stain is the **Gram stain,** which will be discussed in Exercise 9. Other differential stains are the acid-fast stain and the endospore stain.

Materials:
Methylene blue (Loeffler's)
Wash bottle
Bibulous paper
Microscope slides
Bunsen burner
Inoculating loop
China marking pencil
Slide holder (clothespin)

Cultures:
Broth cultures of yeast, *Bacillus megaterium,* and *Staphylococcus aureus*
Nutrient agar slant cultures of *Bacillus megaterium* and *Staphylococcus aureus.*

II. PROCEDURE

1. Prepare smears on slides that have been thoroughly cleaned, or use smears that have been prepared in Exercise 7.
2. Place the heat fixed slide on the staining rack over the sink. See Fig. 8–1.
3. Cover the smear with the staining solution (several different basic dyes can be used—methylene blue, basic fuchsin, malachite green). Allow it to react with the cells for **60 to 90 seconds.**
4. **Rinse** with tap water until all of the staining solution has been removed.
5. **Blot** (do not rub) the slide to remove the water.

From Microbiology Laboratory Manual by Brenda Grafton Wellmeyer. Copyright © 2001 by Kendall/Hunt Publishing Company. Reprinted by permission.

6. **Examine** the slide with the **scanning power (4x), low power (10X), high power (40X),** and **oil immersion (100X)** objectives. The immersion oil should be applied to the surface to the smear. No cover slip is required. (Hint: Focus on the **outer edge** of the smear where you are more likely to find a **monolayer** of cells.)

7. Draw the morphology and arrangement of the cell on the Worksheet.

Do this procedure over the sink.

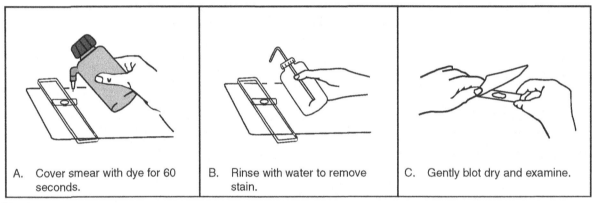

| A. Cover smear with dye for 60 seconds. | B. Rinse with water to remove stain. | C. Gently blot dry and examine. |

FIGURE 8–1

Name _____   **Date** _____

1. In the circles below draw the organisms that you observed using the simple staining technique and write their names below the circles. ˙(Note: Underline each name to designate genus and species).

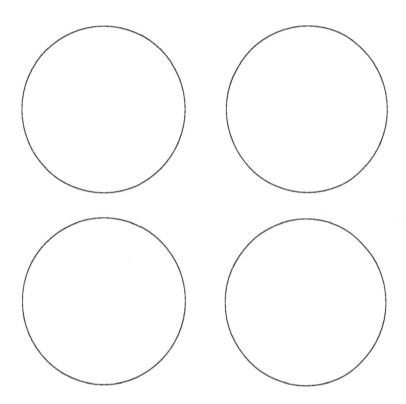

2. Why are basic dyes used for simple staining rather than the acidic dyes which are negatively charged?

3. List three basic dyes that are commonly used in the simple stain.

Exercise 9 THE GRAM STAIN

I. INTRODUCTION

The Gram stain is the most important staining technique used for the identification of unknown bacteria. This procedure reveals the cellular morphology and enables one to differentiate bacteria into two groups according to differences in cell wall chemistry; those that are Gram positive and those that are Gram negative.

The Gram stain technique involves the use of four solutions. The specific reagents used in this procedure are as follows:

1. **Primary stain—Crystal violet,** a basic dye that is absorbed by most bacterial cells.
2. **Mordant—Gram's iodine,** a fixative, which forms an insoluble chemical complex with the crystal violet dye (CV-I complex).
3. **Decolorizer—95% Ethyl alcohol (Ethanol),** removes the lipids (from the cell walls) and dye in the cells.
4. **Secondary Stain—Safranin,** stains those cells that have been decolorized.

The chemical structure of the bacterial cell wall varies. Some species of bacteria possess thicker cell walls composed of multiple layers of peptidoglycan (murein). These bacterial cell walls accept the primary stain, crystal violet, which reacts with the Gram's iodine to form an insoluble precipitate (crystal violet-iodine complex). This complex becomes "fixed" inside the cells and the cells will retain the violet colored complex during subsequent decolorization. These bacteria are classified as **Gram positive.**

The cell walls of some bacterial species consist of a layer of lipopolysaccharides and lipoproteins plus a thin layer of peptidoglycan. The lipid layer occurs on the surface of the cell wall. Removal of this lipid layer by the decolorizer (alcohol) increases the pore size of the cell walls. Consequently, the decolorizer removes the crystal violet-iodine complex and the cells become colorless. These cells readily absorb the pink/red secondary stain, safranin and are classified as **Gram negative.**

The reaction and appearance of the bacterial cells in each step of the Gram stain procedure are summarized in Table 9–1.

TABLE 9–1
The Gram Stain

Gram Stain Solutions	Reaction and Appearance of the Bacterial Cells	
	Gram Positive	**Gram Negative**
1. Crystal Violet	Cells absorb stain. Cells appear violet.	Cells absorb stain. Cells appear violet.
2. Gram's Iodine	Iodine forms complex with crystal violet. Cells remain violet.	Iodine forms complex with crystal violet. Cells remain violet.
3. Ethyl Alcohol (Ethanol)	Cell walls dehydrated. Pores in cell wall shrink. Crystal violet-iodine complex "fixed" within cell. Cells remain violet.	Alcohol removes lipid layer from the cell wall. Large pores occur in cell wall. Alcohol removes crystal violet-iodine complex from the cell. Cells appear colorless.
4. Safranin	Cells will not absorb dye. Cells remain violet.	Cells absorb safranin. Cells appear pink/red.

From *Microbiology Laboratory Manual* by Brenda Grafton Wellmeyer. Copyright © 2001 by Kendall/Hunt Publishing Company. Reprinted by permission.

A number of factors can affect the accuracy of the Gram stain reactions. The critical step involves the use of the **decolorizer,** which can be **over done.** It is possible for a Gram positive bacterium to lose its ability to retain the primary stain and thus appear to be Gram negative. If the **age** of the culture is **beyond 24 hours old,** it may contain some nonviable cells that are beginning to decompose. These cells tend to decolorize more readily causing them to appear Gram negative. Exposing the slide to **excess heat** during fixation may cause the cell walls to rupture causing the cells to over decolorize in the presence of the alcohol. The **reagents** might be **too old** or **improperly mixed.** This may also result in a week or incorrect reaction.

It is also possible for a Gram negative bacterium to appear falsely positive. If the dried **smear is too thick,** the alcohol cannot thoroughly penetrate the smear and decolorize the cells.

This is a precisely standardized procedure and you should follow the directions with care in order to obtain accurate results. Because the Gram stain is the most used of the bacterial stains, it is important that you become proficient in performing this procedure.

Materials:

Slides with heat-fixed smears

Gram-staining solutions and wash bottles

Bibulous paper

Cultures:

Bacillus megaterium (TSA slants)

Escherichia coli (broth)

Staphylococcus aureus (broth)

II. PROCEDURE—GRAM STAIN (FIGURE 9–1)

1. Prepare **thin** smears of the bacteria on a clean slide. Be sure to **heat fix** the smears. Place the slides on the staining rack.
2. Cover the smears with **crystal violet.** Allow it to react for **60 seconds.**
3. **Rinse** with water for 5–10 seconds. Shake the excess water from the slide.
4. Cover the smears with **Grams iodine.** Allow it to react for **60 seconds.**
5. **Rinse** with water 5–10 seconds. Shake the excess water from the slide.
6. Cover the smears with the decolorizer, **ethyl alcohol (ethanol).** Allow it to reach for **10–20 seconds.** Watch the clock carefully so as not to over decolorize.
7. **Rinse** immediately with water for 5–10 seconds. Shake the excess water from the slide.
8. Cover the smears with **safranin.** Allow it to react for **2 minutes.**
9. **Rinse** with water until all dye has been removed. Shake the excess water from the slide.
10. **Blot** (do not rub) the slide dry and examine with your microscope.
11. **Observe** the stained slide using **low power (10X), high power (40X)** and **oil immersion (100X)** objectives. The immersion oil should be placed directly on the smear. Remember to examine an area of the smear where the cells occur in a monolayer. Record your observations on the Worksheet.

   Gram Positive—Violet (Purple)   Gram Negative—Pink/Red

From *Microbiology Laboratory Manual* by Brenda Grafton Wellmeyer. Copyright © 2001 by Kendall/Hunt Publishing Company. Reprinted by permission.

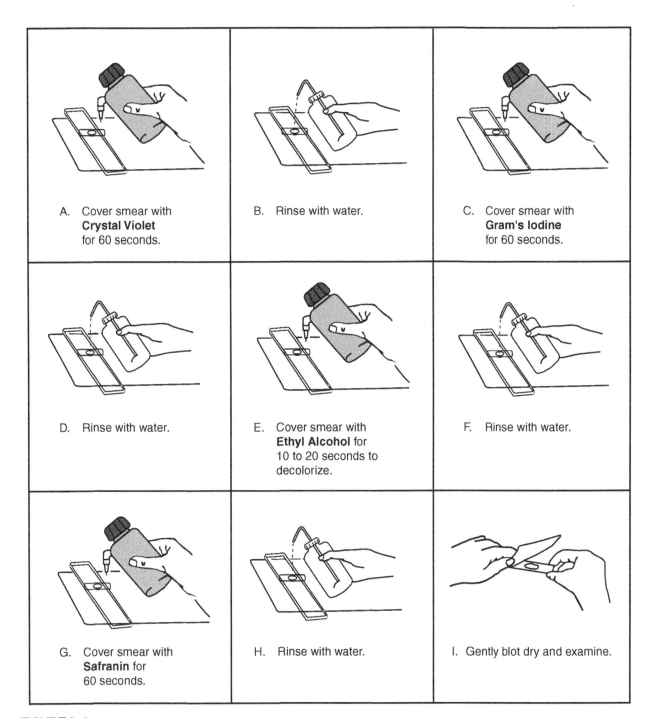

A. Cover smear with **Crystal Violet** for 60 seconds.

B. Rinse with water.

C. Cover smear with **Gram's Iodine** for 60 seconds.

D. Rinse with water.

E. Cover smear with **Ethyl Alcohol** for 10 to 20 seconds to decolorize.

F. Rinse with water.

G. Cover smear with **Safranin** for 60 seconds.

H. Rinse with water.

I. Gently blot dry and examine.

FIGURE 9–1
The Gram Stain Procedure

From *Microbiology Laboratory Manual* by Brenda Grafton Wellmeyer. Copyright © 2001 by Kendall/Hunt Publishing Company. Reprinted by permission.

Name _____   Date _____

Observations: After staining, carefully blot the slide dry. Examine each smear with the oil immersion (100X). Observe the Gram stain reaction and cell morphology of each microorganism. Remember to examine a thin area of your smear (outer edges) where the cells occur in a monolayer. If your results are **incorrect, repeat** the procedure on another set of smears. Record the Gram stain reactions and the microscopic morphology in Table 9–2.

TABLE 9–2
Gram Stains

#	Name of Organism	Gram Stain Reaction	Cellular Morphology

From *Microbiology Laboratory Manual* by Brenda Grafton Wellmeyer. Copyright © 2001 by Kendall/Hunt Publishing Company. Reprinted by permission.

QUESTIONS

1. What would happen if you used methylene blue as a counterstain, instead of safranin?

   _____

   _____

   _____

2. Explain the differences in cell wall structure between Gram-positive and Gram-negative bacteria and the relation of these differences to their color after Gram staining.

   _____

   _____

   _____

3. If you eliminate iodine from the Gram staining procedure, what color would you expect *Staphylococcus aureus* to be after completing the procedure?

   _____

   _____

   _____

From *Microbiology: Laboratory Manual*, 2nd edition by Frank H. Osborne. Copyright © 1994 by Kendall/Hunt Publishing Company. Reprinted by permission.

4. Draw the organisms you observe below. Be sure to label G(+) from G (−) and give the genus and species.

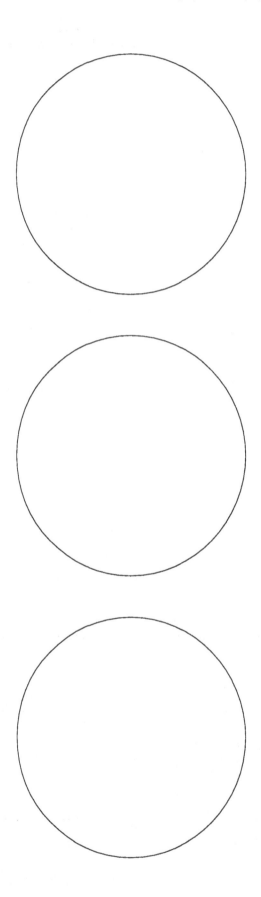

NOTES

Exercise 10   STRUCTURAL STAINS (ENDOSPORE AND CAPSULE)

I. THE ENDOSPORE STAIN

Bacterial endospores are intracellular structures formed by the genera *Bacillus* and *Clostridium*. The size, shape, and location in the cell of the endospore is relatively constant for a given species of endospore-former; this information can be a useful aid in specific identification of some bacteria. Endospores are survival mechanisms produced in response to a depletion of nutrients in the environment. An individual bacterium produces only one endospore in a process called **sporulation.** Sporulation involves a decrease in the activity of the vegetative (actively growing) cell and a marked loss of moisture as all the components of the cell necessary for survival are condensed within a thick wall (spore coat) formed within the cell. Eventually what is left of the vegetative cell disintegrates leaving the free endorspore. It should be noted that this bacterial endospore is not a reproductive structure.

Endospores can withstand relatively harsh environmental conditions. They are relatively resistant to the effects of heat, drying and ultraviolet radiation. Endospores are also resistant to many chemicals, including a number of commonly used disinfectants. They may persist for many years in soil—some spore of *Bacillus anthracis* are known to have remained viable for more than one hundred years. There are some published accounts of endospores being germinated many hundreds of years after their initial production! Germination of an endospore when nutritional conditions are again favorable produces a single bacterium identical to the vegetative cell that originally produced it.

Both genera, *Bacillus* and *Clostridium*, contain important human pathogens. *Bacillus anthracis* causes the disease anthrax, while *Colstridium botulinum* causes botulism. In both cases the endospore plays a significant role in the disease process. In respiratory anthrax it is the inhalation of the endospores that initiates this very serious disease. In botulism, more properly called an intoxication than an infection, endospores that are not destroyed by the heat applied during food canning processes find themselves in a nutritionally rich, anaerobic environment. In such an environment the endospores readily germinate and the resulting vegetative cells produce the botulinum toxin. Consuming this preformed toxin results in the life-threatening symptoms of botulism. It should be noted that infant botulism is most probably caused when a child under the age of 12 months ingests endospores and they germinate within the child's intestinal tract. The resulting vegetative cells elaborate the toxin and the child becomes ill. The Centers for Disease Control in Atlanta have recommended that honey, the confirmed source of endospores in many cases of infant botulism, not be fed to children less than one year old.

Because the endospore is highly impermeable it is difficult to stain. Endospores may be seen as clear areas within Gram stained bacteria. Free endospores however, cannot be seen unless a special staining procedure is used. The most common procedure uses heat to drive malachite green into the spore coat. Safranin is often used as a counterstain that allows visualization of the vegetative cell as well.

From *Fundamental Microbiology for the Health Care Sciences,* Fourth Edition by Frank A. Hartley, Walter Hoeksema, and Michael Ryan. Copyright © 2001 by Kendall/Hunt Publishing Company. Reprinted by permission.

Materials:

Electric hot plate

Small beaker (25 mL)

Spore Staining solutions: 5% malachite green and safranin

Cultures:

24-96 hour TSA Slants of *Bacillus megaterium*.

II. PROCEDURE

1. Place approximately 350–400 ml of tap water into the staining beaker. Place it on the hot plate to begin to boil. See Figure 10–1.
2. Prepare a smear of *B. megaterium* using a loop of water.
3. After air drying (about 5 minutes), heat—fix the slide.
4. Place the slide on the staining rack, cover with a paper towel strip and flood the paper with malachite green stain.
5. Steam the slide for 4 minutes, adding stain to prevent drying.
6. Carefully remove the slide, cool and wash with water from the wash bottle.
7. Counterstain with safranin for 1 minute.
8. Wash off excessive stain, blot dry and examine under oil immersion.
9. Draw and label the endospores and vegetative cells of *B. megaterium* under the data section. Indicate the particular colors observed. *Note: Malachite green stains the endospore and safranin stains the vegetative portion of the cell. Therefore, you should see a green endospore contained in a pink sporangium.

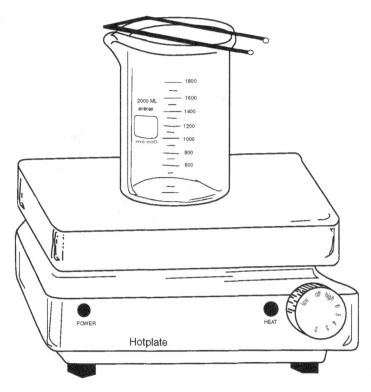

FIGURE 10–1
Hot Plate and Staining Apparatus

III. FURTHER STUDIES: STRUCTURAL STAIN: THE PAYNE SPORE STAIN METHOD (VARIATION OF THE ENDOSOPORE STAIN)

IV. PROCEDURE

1. Prepare a smear and cover the smear with malachite green and drain the excess fluid from the slide.
2. Let the dye set for 5 seconds.
3. Steam the slide in a slide tray for 5 minutes over a hotplate. The glass slide tray holder should have at least 2 ml of distilled water or at least enough to cover the bottom of the slide tray. *Note: The hotplate should be set between low and medium heat.
4. Remove the slide from the tray and let the slide cool, then rinse with water for 30 seconds.
5. Counter stain with safranin for 30 seconds.
6. Rinse with water.
7. Blot dry with bibulous paper and examine the slide using the oil immersion lens.

Name _____   **Date** _____

1. Draw your results using colored pencils.

2. Discuss the differences in oxygen tolerance between the genus *Bacillus* and the genus *Clostridium*. Use Bergey's Manual.

3. How long can spores survive?

4. What conditions favor spore production in endospore forming bacteria?

5. Why is it so difficult to stain endospores?

6. What would be the best way to destroy an endospore?

Adapted from *Microbiology: Laboratory Manual*, 2nd edition by Frank H. Osborne. Published by Kendall/Hunt Publishing Company.

7. Draw your observations below. Identify the spores and the genus and species of the bacteria.

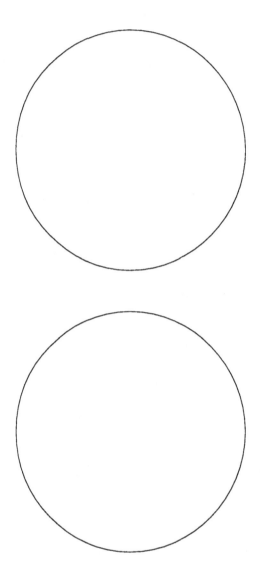

STRUCTURAL STAIN: THE CAPSULE STAIN

A capsule is a gelatin-like layer that surrounds some bacteria and fungi. It is produced and secreted by the microbe. The capsular material (often polysaccharide in bacteria) may be present in large amounts rendering the capsule visible with special staining procedures, or the material may be present in very small amounts detectable only through the use of serological techniques. The presence of a capsule surrounding a bacterium is often discernable as the colonies of the bacterium grow on solid culture media. For example, pneumococci with capsules produce colonies that are "watery" and mucoid (sometimes called smooth) whereas pneumococci without capsules produce colonies that are "dry" and non-mucoid (sometimes called rough).

Microbes that possess capsules are often resistant to host defense mechanisms, especially to phagocytosis. In this context the capsule can be considered a virulence factor. However, the presence of a capsule is not necessary for the normal survival of a microbe. Some pneumococci are not encapsulated and, although they do not cause pneumonia because they are readily phagocytizeed, they are able to thrive in environments lacking such defense mechanisms, e.g., a culture dish in the laboratory.

Among the microbes that possess capsules are the bacteria *Streptococcus pneumonia* (the pneumococcus), *Neisseria meningitides* and *Klebsiella pneumoniae* and the fungus *Cryptococcus neoformans*.

Simple staining reagents are not very effective for visualizing the capsule. Negative staining reagents that color the background but only outline microbial surface structures are required. One such reagent is India ink. Microscopic examination of a suspension of an encapsulated microbe in a solution if India ink reveals the capsules as clear areas in a black background. The presumptive identity of the causative agent of a disease like meningitis that requires prompt therapeutic decisions can often be achieved by mixing a sample of spinal fluid with India ink. The characteristic morphology and cell-to-cell arrangement of one common cause of meningitis is readily seen in this way though only the capsules are visualized—again as clear areas against a black background.

In this lab exercise you will couple a negative stain with a direct stain to visualize not only the microbial capsule, but also the microbe within.

Materials:

36–48 hour mild culture of *Klebsiella pneumoniae, Streptococcus pneumoniae*

Aqueous Crystal violet

Copper sulfate (20%)

3% salt solution

Safranin

Nigrosine or Congo red

PROCEDURES

1. Prepare a smear of *Klebsiella pneumoniae or Streptococcus pneumonia.* Do not heat fix; you may destroy the capsules.
2. Stain with aqueous crystal violet for 2 minutes.
3. Use no water. Wash the slide with 20% copper sulfate
4. Air dry and observe.

1. Stain the smear with aqueous crystal violet for 2 minutes.

2. Wash the slide with 20% copper sulfate.

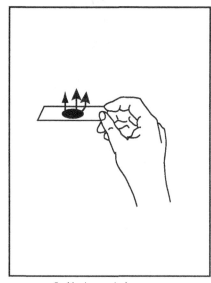

3. Air dry and observe.

FIGURE 10–2

Material adapted from *Microbiology: Laboratory Manual* by Frank H. Osborne and *Microbiology Laboratory Manual* by Brenda Grafton Wellmeyer. Published by Kendall/Hunt Publishing Company.

THE STROBERT-PAYNE CAPSULE STAINING METHOD (VARIATION OF THE CAPSULE STAIN)

CAPSULES

The ability to form a capsule is genetically determined, but the size of the capsule is influenced by the medium on which the bacterium is growing. In this procedure, a salt solution will be added to shrink the bacterial cell and to give the capsules a larger appearance when examined under the microscope.

Materials:

Klebsiella pneumoniae (48-hour)

Streptococcus pneumoniae (24-hour)

PROCEDURE

1. Add A Drop of 3% salt solution to the microscope slide.
2. Suspend a small amount of the organism from the petri plate in the salt solution and mix thoroughly for 1 minute.
3. Add 1 drop of safrainin for 1–2 minutes and mix.
4. Add 1 drop of nigrosine for 1 minute and mix.
5. Using a sterile inoculating loop, spread the suspension over the slide.
6. Excess dye should be absorbed by using a paper towel or bibulous paper, by holding the microscope slide horizontally at a right angle.
7. Congo red can also be used in lieu of nigrosine.

QUESTIONS

1. What was the purpose of using a salt solution for this procedure?
2. Draw a diagram of the cells that you have observed.

_____

_____

Name _____   **Date** _____

1. Describe a clinical application of the capsule stain.

2. Encapsulated *Streptococcus pneumoniae* can cause pneumonia. Unencapsulated S. pneumoniae does not cause pneumonia. Explain.

TRUE OR FALSE QUESTIONS

1. Bacterial capsules are easily observed using simple staining techniques. _____
2. Fungal microbes cannot produce capsules. _____
3. The capsule is secreted onto the microbe's external surface by the microbe. _____
4. Bacterial capsules do not aid the bacteria in causing human disease. _____
5. The capsule stain does not, in fact, stain the capsule. _____
6. Crystal violet is the negative staining reagent in the capsule stain. _____
7. Microbes must first be grown on artificial medium before their capsules can be observed using the capsule stain. _____
8. As part of the capsule stain as done in lab the cell wall of *Klebsiella pneumoniae* is stained. _____
9. A capsule is a gelatin-like material that surrounds some bacteria and fungi. _____
10. Bacterial capsules are usually composed of polysaccharides. _____
11. A microbe that normally possess a capsule will die if the capsule is removed. _____

NOTES

Exercise 11   ACID-FAST STAIN

I. INTRODUCTION

The Acid-fast stain is a differential stain used to distinguish members of the genera *Mycobacterium* and *Nocardia* from other organisms which are not acid-fast. These bacteria have cell walls which contain waxes and mycolic acids of high molecular weight. Members of the genus *Corynebacterium* share this wall structure but are not acid-fast. This stain is critical when tuberculosis or leprosy are suspected.

The acid-fast organisms will take on a characteristic red color when stained with hot carbol fuchsin. This red color is not removed by an alcohol solution of acid, hence the name "acid-fast". The acid alcohol solution will remove the red color from cells of other genera. Methylene blue is used as a counterstain to visualize non-acid-fast organisms.

Materials:

Mycobacterium smegmatis
Carbol fuchsin
Acid Alcohol
Methylene blue
Paper towel or filter paper
Microscope slide
Beaker
Screen
Tripod
Bunsen burner

II. PROCEDURE: ACID-FAST STAINING.

1. Set up the apparatus as shown in the drawing in Figure 11–1. Use a can with a small enough diameter to hold a slide on it. The purpose of the screen is to keep the can from falling through the hole in the tripod, not to assure even distribution of heat as is done in chemistry with glass containers.

FIGURE 11–1
Apparatus for Acid-Fast Staining

2. Prepare a smear of *Mycobacterium smegmatis* that has been growing on a nutrient agar slant for a few days. As a control, mix in cells of some other organism; anything will do. Air dry and heat fix the slide.

3. Place the slide over the can of boiling water and cover the smear with a 1" square of paper towel or filter paper. Drop sufficient carbol fuchsin on the paper to keep the center of the paper moist. Boil for 5 minutes adding more stain as necessary. After 5 minutes, remove the slide, take the paper off and wash with water.

4. Decolorize with acid-alcohol for 10–20 seconds. Immediately wash with water.

5. Counter stain with methylene blue for 30 seconds. Wash with water, dry and observe the slide.

NOTES

Name _____   **Date** _____

1. Draw the appearance of your correctly stained *Mycobacterium* preparation with colored pencils.

2. Describe the usual mycelial growth patterns in the genus *Nocardia*.

3. Give a disease caused by each organism listed below.
   Mycobacterium ————————————————————————
   Nocarida ————————————————————————
   Corynebacterium ————————————————————————

Exercise 12   MOTILITY DETERMINATION

I. INTRODUCTION

A wide variety of bacteria exhibit motility—the ability to move from place to place. Motility generally results from the presence of flagella. Flagella are thin structures that are much smaller than the limit of resolution of the oil immersion lens. Evidence of motility is seen when bacteria move long distances in single direction. This can be contrasted with mere bouncing around in place, which is called Brownian Movement. Motile organisms generally move at a high rate of speed relative to the non-motile objects in the field.

The hanging drop slide technique is used to observe living bacteria and to determine if they are motile. In this lab, a culture of motile bacteria will be compared with a culture of non-motile bacteria.

Materials:

Depression slide

Cover glass

Trypsticase soy broth cultures of *Micrococcus luteus* and *Proteus vulgaris*

Incoculating loop

Bunsen burner

Vaseline

II. PROCEDURE: HANGING DROP SLIDE.

1. Prepare a depression slide by depositing four small drops of glycerol at the corners of the depression. These will suspend coverslip above the depression. See the first part of Figure 12–1.

2. With a loop, remove a drop of culture to be studied and place it in the center of a *cover slip* (do not place it in the center of the depression slide). Make sure that you have a good sized drop—small ones dry out too fast.

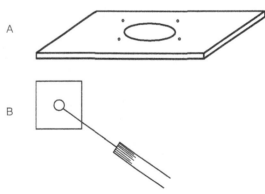

FIGURE 12–1
Steps A and B Of the Hanging Drop

From *Microbiology: Laboratory Manual*, 2nd Edition by Frank H. Osborne. Copyright © 1994 by Kendall/Hunt Publishing Company. Reprinted by permission.

3. Invert the depression slide over the cover slip, center it, and touch it to the cover slip. The cover slip should adhere to the depression slide.

4. Invert the depression slide so that the coverslip is now on top. You should now have the drop of culture hanging from the cover slip over the depression.

5. Place a drop of oil on the cover slip over the spot where the drop is. This will help you to center the preparation.

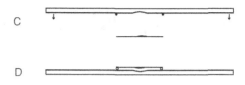

FIGURE 12–1
Steps C and D Of the Hanging Drop

NOTES AND OBSERVATIONS

QUESTIONS

1. How do flagella enable bacteria to swim?

_____

_____

_____

_____

_____

_____

_____

2. List the differences in structure between bacterial flagella and eukeryotic flagella.

_____

_____

_____

_____

_____

_____

3. Explain the difference between Brownian movement and true motility of bacterial cells based on your observations.

_____

_____

_____

_____

_____

_____

4. What causes Brownian movement?

_____

_____

_____

_____

_____

_____

CULTURE TECHNIQUES

Exercise 13   WORKING WITH MEDIA

I. INTRODUCTION

Microorganisms in general and bacteria in particular are dependent upon many factors for growth, both in their natural habitats and in the laboratory. The nutrient preparation used for bacterial growth are called **media** (singular, **medium**). How bacteria grow and the rates at which they grow in these media are dependent upon many physical and chemical factors, some of which are described below.

Physical factors affecting bacterial growth include pH, temperature, oxygen, moisture, and osmotic pressure. The acidity or alkalinity of a particular medium affects the **pH** of the medium. Generally, bacteria prefer to grow at or around a neutral pH, but exceptions do exist. Bacteria are able to grow over a wide range of **temperatures.** Each bacterium has its own optimum growth temperature, although it may be able to grow at temperatures above or below this temperature. The presence or absence of **oxygen** in a growth medium greatly affects the growth of bacteria. Some bacteria cannot grow in the presence of oxygen, while others must have oxygen present or they will die. All actively growing bacteria must have **moisture** (water) present in their immediate environment. However, certain bacteria produce spores which can survive for years in dry environments. Finally, the **osmotic pressure** between the bacterial cytoplasm and the environment must be approximately equal for most microorganisms. Again, exceptions to this rule do exist. We will study each of these physical factors in greater detail in Exercise 16–20.

As with human beings, the growth of microorganisms is affected by chemical/nutritional factors. Factors necessary for bacterial growth include **carbon, nitrogen, sulfur, phosphorus, trace elements,** and **vitamins.** Commonly, these elements are provided to bacteria in the growth medium. Some bacteria, however, are considered **fastidious;** that is, they have special nutritional requirements that go above those just described. Such organisms can be quite difficult to grow in the laboratory.

Microorganisms may be cultured in two types of media, **complex media** or **synthetic (chemically defined) media. Complex media,** used in routine laboratory work, are rich in basic nutrients, such as amino acids, sugars, vitamins and minerals. Most complex media are composed of yeast extracts, meat extracts, peptones (partially digested proteins), mild protein (casein) or gelatin. The exact chemical composition varies slightly from batch to batch. **Synthetic (chemically defined) media** are prepared so that the definite chemical composition is known. These media are composed of exact amounts of pure chemicals rather than materials such as meat extracts. Since all components are known both qualitatively and quantitatively their composition is always the same.

In addition to being defined (synthetic) or complex, culture media may also be classified on the basis of its chemical composition, use or function. The main types of culture media include:

1. **General Purpose** or **Simple Media**

   Nutrient broth is a liquid complex medium containing basic nutrients require for microbial growth. **Nutrient agar** is nutrient broth containing 1.5% to 2% agar. General purpose media will support the growth of a large variety of microorganisms and are the most commonly used type of bacteriological media.

2. **Enriched Media**

   General purpose media will not support the growth of a number of microorganisms (such as the pathogens) with the special dietary requirements. Enriched media is supplemented with whole blood, serum, vitamins, sugars, amino acids, etc. to promote the growth of these microorganisms.

3. **Selective Media**

   Selective media contain chemicals that permit the growth of only one particular kind (species) of microorganism while limiting or inhibiting the growth of others. This type of medium is used when trying to isolate a specific microorganism from a mixed population.

From *Exercises for the Microbiology Laboratory* by Elizabeth Fish McPherson. Copyright © 2001 by Elizabeth Fish McPherson. Reprinted by permission of Kendall/Hunt Publishing Company

4. **Differential Media**

These media are composed of specific nutrients that enable the microbiologist to distinguish between microorganisms n the basis of their biochemical activities. For example, some bacteria may be identified by cultivation in media containing a particular sugar and an indicator. The indicator will change colors to indicate a change in pH (hydrogen ion concentration) resulting from fermentation. Thus, a color change in the medium following inoculation and incubation indicates a positive reaction.

Microbiological media can be used in three forms, **liquid (broth), semisolid,** and **solid.** Essentially, all three forms are identical, with the exception that the semisolid and solid media contain varying amounts of the solidifying agent, **agar,** a polysaccharide derived from marine algae. Agar is an ideal solidifying agent because only a few organisms degrade it, it does not melt until the temperature reaches 95°C, and after being melted, it remains in the liquid state until it has cooled to about 40°C. At this temperature, the medium is cool enough to allow the addition of nutrients and living organisms which might be destroyed by higher temperatures. While in the liquefied state, media containing agar can be poured into either a test tube or a **petri plate.** If the liquefied media is allowed to solidify in a slanted position, the resulting media is called an **agar slant.** If the liquefied media is allowed to solidify with the tube remaining upright, the resulting media is named an **agar deep.** If the liquefied media is poured into a large test tube and is intended for pouring into a petri plate at a later date, the media is referred to as an **agar pour.** Liquefied media poured into a petri plate in order to create a solid surface for growing bacteria is called an **agar plate.**

After preparation and sterilization, all media should be incubated or held at room temperature for at least 24 hours in order to check for growth of contaminants before it is used.

Materials:

100 ml hot liquid nutrient agar

4 sterile Petri dishes

Marking pen

Oven mitts or rubberized "hot hands"

II. PROCEDURE: PREPARATION OF AGAR PLATES

Each student will receive one flask containing 100 ml. of **hot** liquid nutrient agar. Before lab, the dehydrated medium was weighed and placed in the flask. 100ml. of deionized water was added to the flask. The flask was swirled to suspend the powder in the water and a plug was placed in the mouth of the flask. The media was sterilized in the autoclave. The sterile liquid medium must be poured into sterile petri dishes **without contaminating** the medium.

1. Sanitize your desktop and obtain a Bunsen burner.
2. Carefully remove four (4) sterile petri dishes from the plastic bag **without** separating the bottom and the lid. The plates are stacked upside down in the bag to make it easier to remove the two together.
3. Initial the **bottom** of each plate.
4. Line the plates up near the edge of your desktop.
5. Obtain a flask of sterile liquid medium. Use several sheets of paper towel or the rubberized "hot hands" to carry the hot flask. Grasp the neck of the flask with one hand and support the bottom of the flask with your other hand.
6. **Gently swirl** the flask to mix.
7. You may use the "hot hands" to hold the flask. If using paper towels, place several sheets of paper towels around the bottom of the hot flask to protect your hands (see figure 13–1). Do not wrap the towel around the mouth of the flask because it may catch fire while flaming the mouth of the flask.

FIGURE 13–1

8. Pick up the flask, holding it at an **angle** (see figure 13–2) so that airborne organisms cannot drop directly into the sterile medium when the plug is removed.

9. Twist the plug several times to prevent sticking. Remove the plug and set it aside. After removing the plug, continue to hold your flask at an **angle** at all times.

10. Rotate the mouth of the flask through the flame of your Bunsen burner several times to kill organisms around the lip of the flask (figure 13–3). This is referred to as **flaming** the flask.

11. Carefully raise the lid of a perti dish only high enough to pour the liquid agar into the dish. Fill with enough liquid agar to form a layer approximately one fourth (¼) inch deep in the bottom of the dish (figure 13.4).

12. Continue pouring liquid agar into the three (3) remaining dishes. It is advisable to flame the mouth of the flask once more between the second and the third dishes.

13. Allow the agar to cool without disturbing the plates. The agar will solidify when it reaches 40°C – 45°C.

14. After the agar solidifies, the petri dishes should be **inverted** for storage. This prevents the moisture that accumulates on the lid from dropping onto the surface of the agar. Excess moisture on the agar's surface causes swarming of bacterial growth preventing isolation.

15. The freshly poured agar plates should be held at room temperature for a minimum of 24 hours to check for contamination.

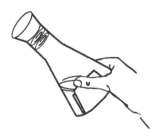

FIGURE 13–2

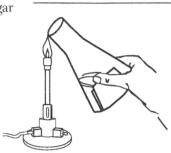

FIGURE 13–3

FIGURE 13–4

III. LABORATORY ACTIVITIES

1. Demonstrations:
   a. A display of dehydrated media.
   b. A display of various types of prepared media.
   c. The autoclave.
2. Preparation of Nutrient Agar Plates:

Each student will prepare **four** (4) nutrient plates following the directions on pages 18 and 19. After the agar plates have solidified, turn them **upside down** and place them in your lab drawer until the next class period.

IV. QUESTIONS

1. List four characteristics of agar that make it useful as a solidifying agent for microbiological media.

   a.

   b.

   c.

   d.

2. Define sterilization.

   _____

   _____

3. List the standard conditions for steam sterilization.

   _____

   _____

4. What is the difference between selective and differential media?

   _____

   _____

5. Why is a medium sterilized prior to its use in the microbiology lab?

   _____

   _____

6. What is an autoclave?

   _____

   _____

7. List the physical and chemical factors affecting bacterial growth.

   _____

   _____

8. What would probably happen to an agar-based solid medium that was inoculated with a microbe that utilized agar as a nutritional substrate?

   _____

   _____

NOTES

Exercise 14   ISOLATING PURE CULTURES

INTRODUCTION

The process of successfully separating an individual microorganism from a mixed population is called **isolation.** A culture containing many kinds of microbes is known as a **mixed culture.** One that contains a single type of microorganism is called a **pure culture.** The identification and study of the biochemical properties of a microbe requires the use of a pure culture.

In order to obtain a pure culture, the microbiologist must have an effective method of isolating a single type of microorganism from a mixed culture. This may be accomplished by diluting the bacterial population contained in the inoculum so that the individual cells are spread out or separated either within a solid medium or on its surface.

As bacteria multiply on (or in) solid media, visible clumps of cells are formed. These clumps are known as **colonies.** Each colony consists of a mass of cells that arises from a single bacterial cell. Colonies produced by different microbial species vary in observable characteristics (size, shape, color, texture). This is referred to as **colony morphology.** Distinguishing morphological characteristics produced by different species of bacteria provide a valuable aid in differentiation and identification. Since all microorganisms in a single colony are "descendants" of a single cell, a **pure culture** can be obtained by aseptically transferring a small inoculum from an isolated colony to a fresh medium.

Two methods used to attain isolation are the **streak plate** method and the **pour plate** method. The **streak plate** method consists of aseptically spreading the inoculum over the surface of a sterile agar plate. The large population of cells will be gradually reduced so that individual cells are distributed over the agar's surface. Each cell will produce a colony. This method of isolation is one of the most widely used techniques in the bacteriology laboratory. See Figure 14–1.

From *Microbiology Laboratory Manual* by Brenda Grafton Wellmeyer. Copyright © 2001 by Kendall/Hunt Publishing Company. Reprinted by permission.

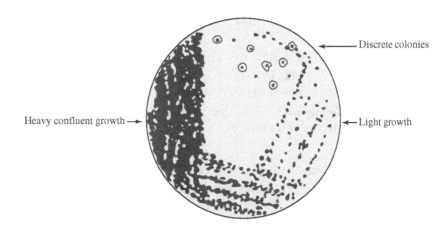

Discrete colonies

Heavy confluent growth →

← Light growth

FIGURE 14–1
Streak Inoculation

From *Microbiology: Laboratory Manual* by Jake H. Barnes and Randall M. Brand. Copyright © 1995 by Jake H. Barnes and Randall M. Brand. Reprinted by permission of Kendall/Hunt Publishing Company.

In the **pour plate method,** the inoculum is aseptically mixed (or diluted) in sterile liquid agar medium and poured into an empty sterile petri dish. As the agar cools it solidifies. Colonies develop within and on the surface of the agar plate. This method may be used when it is desirable to determine the size of the bacterial population in the original sample. Several dilutions may be made in the liquid media before the plates are poured. Following incubation the number of colonies growing in the culture can be counted and the total number of bacteria can be calculated.

Once media has been inoculated it is held at a temperature favorable for maximum multiplication of that microorganism. This is called **incubation.** Each species of microorganism has a preferred range of temperature at which the most rapid rate of growth occurs. The temperature at which an microorganism grows best is called the **optimum growth temperature.**

From *Microbiology: Laboratory Manual* by Jake H. Barnes and Randall M. Brand. Copyright © 1995 by Jake H. Barnes and Randall M. Brand. Reprinted by permission of Kendall/Hunt Publishing Company.

I. STREAK PLATE METHOD: THE QUADRANT STREAK

Materials:

1 sterile Petri dish

Wax pencil or Permanent marking pen

2 nutrient agar plates

Bunsen burner

Wire loop

Cultures:

1 mixed culture of *Serratia marcescens, Escherichia coli,* and *Micrococcus luteus*

Procedure:

A. Dry Run Using Empty Petri Dish and Marking Pen:

1. Each student should obtain a Petri dish and a marking pen.
2. Turn the Petri dish upside down, and mark it on the bottom with the marking pen as shown in Fig. 14–2 A.
3. Next, turn the unopened empty dish right side up.
4. With the Petri dish lying on the table in front of you, **carefully** open the lid slightly, but keep the lid **over** the dish. (This will protect the sterile medium from contamination by bacteria or fungi.)
5. With the marking pen, draw, on inside of the dish, the lines that are shown in Fig. 14–2 B. This represents the first set of streaks of bacteria you will be using for the Quadrant Streak method.
6. Close the lid of the dish.
7. Rotate the dish, counterclockwise, one quarter turn.
8. Open the lid again, and draw the lines that are shown in Fig. 14–2 C.
9. Close the lid.
10. Rotate the dish again, and continue "streaking" the plate, as described above, until you have completed the examples shown in Fig. 14–2 D.
11. Have your streaking patterns checked by your instructor.

B. Streaking for Isolation Using Living Bacteria:

1. Each student should obtain 2 nutrient agar plates, a wire loop, and 1 mixed culture of bacteria.
2. Mark the bottom of the Petri dish with the correct information, as shown in Fig. 14–2 A.

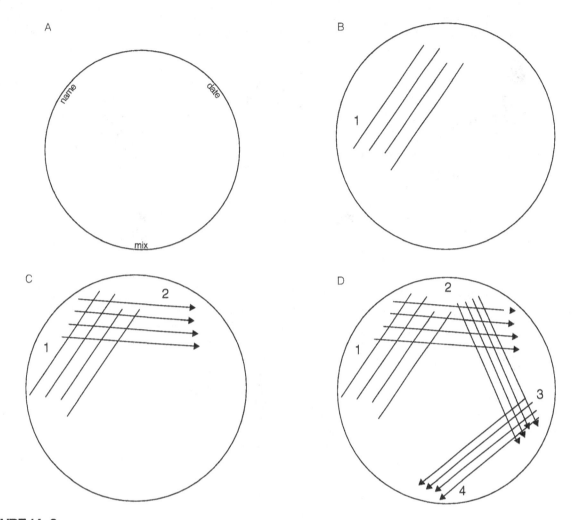

FIGURE 14–2

From *Microbiology: Laboratory Manual* by Jake H. Barnes and Randall M. Brand. Copyright © 1995 by Jake H. Barnes and Randall M. Brand. Reprinted by permission of Kendall/Hunt Publishing Company.

3. Using a sterile wire loop, aseptically obtain a loopful of broth from the mixed culture. **(You will not need the culture for the rest of the exercise.)** The aseptic transfer of a culture from broth to a Petri plate is shown in Fig. 14–3.

4. Open the lid carefully, and make 4 streaks in area 1 (Fig. 14–2 B). Try to keep the lid of the Petri dish over the agar as much as possible. Hold the loop as flat as possible, to prevent the tearing of the agar.

5. Close the lid.

6. Flame the loop and let it cool for 10 seconds.

7. Rotate the Petri dish, counterclockwise, one quarter turn.

8. Open the lid again, and pass the loop through area 1, across the previous streaks, and continue streaking through area 2 (Fig. 14–2 C).

9. Close the lid. Flame the loop. Repeat steps 6, 7, and 8 for streaking areas 3 and 4 (Fig. 14–2 D). Make sure you flame the loop between streaking each area. Be careful that your last 4 streaks do not touch your first streaks in area 1.

10. When you are finished, sterilize your loop.

11. Repeat the above technique using your second nutrient agar plate.

12. Incubate the plates at 25°C for 7 days.

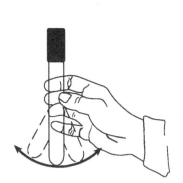

1 Shake the culture tube from side to side to suspend organisms. Do not moisten cap on tube.

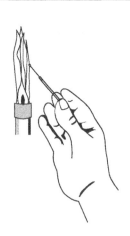

2 Heat the loop and wire to red-hot. Flame the handle slightly also.

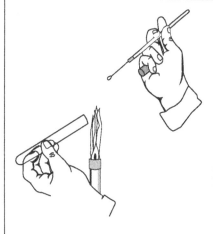

3 Remove the cap and flame the neck of the tube. Do not place the cap down on the table.

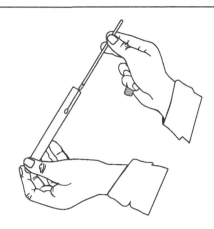

4 After allowing the loop to cool for at least 5 seconds, remove a loopful of organisms. Avoid touching the sides of the tube.

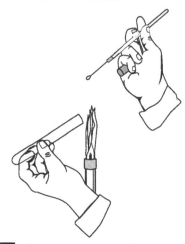

5 Flame the mouth of the culture tube again.

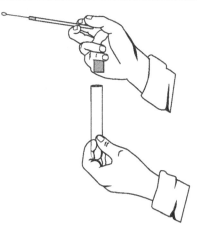

6 Return the cap to the tube and place the tube in a test-tube rack.

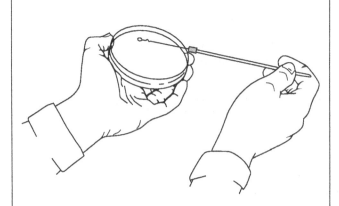

7 Streak the plate, holding it as shown. Do not gouge into the medium with the loop.

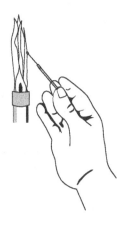

8 Flame the loop before placing it down.

FIGURE 14–3

II. POUR PLATE METHOD

Materials:

3 sterile Petri dishes   Beaker with water (1/3 full)

3 nutrient agar pours   Wire loop and marking pen

Electric hot plate

Cultures:

1 mixed culture (from Streak Plate Method, above)

Procedure:

1. Label the agar pours I, II, and III. Also, label the three Petri dishes I, II, and III, and include other appropriate information.

2. Melt the 3 nutrient agar pours in a boiling water bath.

3. Cool the agar to 50°C. You will have to work quickly, at this point, so that your agar does not solidify. Read the steps below, to make sure you are prepared to perform the task in a timely fashion.

4. Follow the basic techniques that are shown in Fig. 14–4.

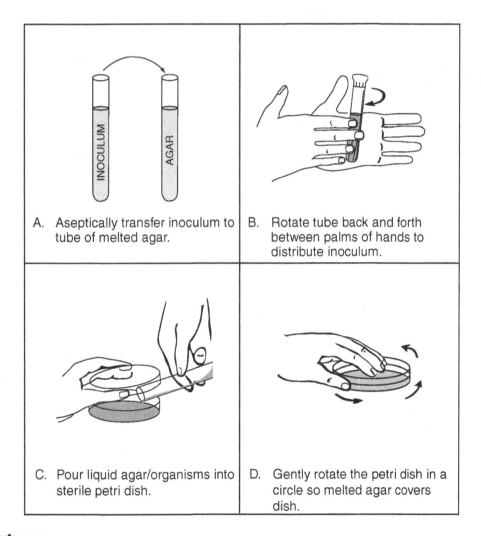

A. Aseptically transfer inoculum to tube of melted agar.

B. Rotate tube back and forth between palms of hands to distribute inoculum.

C. Pour liquid agar/organisms into sterile petri dish.

D. Gently rotate the petri dish in a circle so melted agar covers dish.

FIGURE 14–4

5. Using aseptic techniques, inoculate tube I with a single loopful of the mixed culture (inoculum).

6. Carefully mix the tube by rotating it back and forth in the palms of your hands.

7. Immediately, using aseptic techniques, remove one loopful from tube I and transfer it to tube II.

8. Place Tube I back into the 50°C water bath.

9. Carefully mix Tube II.

10. Immediately, using aseptic technique, Remove one loopful from Tube II and transfer it to Tube III.

11. Place Tube II back in the water bath.

12. Carefully mix Tube III.

13. Place Tube III back in the water bath.

14. Using aseptic technique, pout the contents of Tube I into a sterile Petri dish. You need to work quickly, so the agar does not solidify. Gently rotate the plate to spread the agar evenly over the bottom of the plate.

15. Repeat step 14 for Tubes II and III.

16. Incubate the plates at 25°C for 24–48 hours.

III. EVALUATION OF THE TWO METHODS

After 24–48 hours, record your results for the Streak Plate and Pour Plate methods on the Worksheet at the end of Exercise 14.

IV. OBTAINING PURE CULTURES

After using the Streak Plate method of isolation, the next step in obtaining a pure culture is to transfer the organisms from a single colony, on the Petri dish, to a tube of nutrient broth, or a nutrient agar slant.

Materials:

3 nutrient agar slants

Wire loop or needle

Bunsen burner

Procedure:

1. Label one agar slant *E. coli*, another *S. marcescens*, and the third *M. luteus*.

2. Select an isolated **WHITE** colony (contains *E. coli*)

3. Take a sample of the colony with your wire loop.

4. In the tube labeled *E. coli*, streak the agar slant by placing the loop near the bottom of the slant, and gently drawing it up over the surface of the agar. One streak is enough.

5. Repeat the inoculating procedure for the other two bacteria. The **RED** colony is *S. marcescens*, and the **YELLOW** colony is *M. luteus*.

6. Incubate for 5 to 7 days, 25°C.

7. Evaluate each slant to see if you obtained a pure culture. This can be done visually, and by performing a Gram stain for each of the subcultures.

8. Record your result on the Worksheet.

Name _____   **Date** _____

WORKSHEET

1. Sketch the appearance of your streak plates and pour plate.

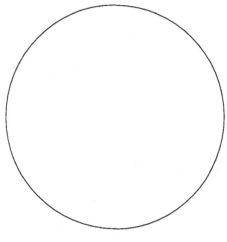
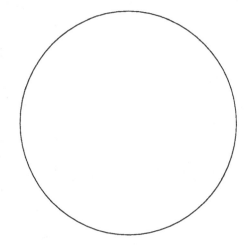

Streak plate method   Pour plate method

2. Which technique, the Quadrant or Radiant method, proved to be the best method for you in the isolation of a pure culture?

3. Record observations of your streak plate.

	Number of Isolated Colonies			
	white	red	yellow	total number
Plate 1				
Plate 2				

4. Subculture Evaluation

   With colored pencils, sketch the appearance of the growth on the slant diagrams below. Also, draw a few cells of each organism as revealed by Gram staining in the adjacent circle.

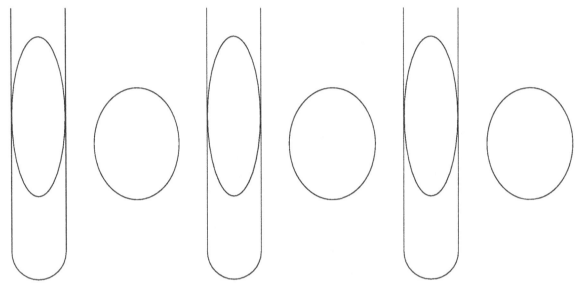

Serratia marcescens   Micrococcus luteus   Escherichia coli

5. Do you think you have pure cultures of each organism on the slants?_____

   Can you be absolutely sure by studying its microscopic appearance? _____

   _____

   Explain: _____

QUESTIONS

1. Describe three circumstances in which pure cultures are essential for the microbiologist.

   a.

   b.

   c.

2. Define pure culture.

   _____

3. Why must the nutrient agar be cooled to 50°C before inoculation and pouring?

   _____

4. Why must you invert plates during incubation?

   _____

5. Why must the loop be flamed before entering a culture to be isolated?

   _____

NOTES

Exercise 15   BACTERIAL POPULATION COUNTS

I. INTRODUCTION

In many applications, a microbiologist must determine the number of bacterial cells in a particular sample. For example, suppose that you wish to determine the effectiveness of boiling in removing bacteria from a water sample. You would need to count the concentration of viable bacteria on the water sample before and after the treatment.

A microscopic count is not usually informative, because both live and dead cells as well as small debris could be counted. Instead, an estimate of the number of viable cells can be determined by applying a measured quantity of the sample to a petri dish and counting the number of colonies that form. The number obtained by this approach is called a viable cell count, because only living cells are capable of reproducing and forming a colony.

Ideally, a colony is formed by the reproduction of a single cell that is separated from other cells. However, in reality this is probably very rare. Instead, most bacterial cells are arranged in groups: chains, clusters, tetrads, etc. For this reason, the number given by viable cell count methods is not called "cell." It is called "colony forming units (CFU)." Within a certain range of bacterial numbers, the number of colony forming units in a sample is directly proportional to the number of viable bacterial cells. Because the multiple fluid transfers involved in the viable count introduce the possibility of contamination, fewer than 30 colonies are not counted. Because large numbers of colonies are difficult to keep track of as they are counted, greater than 300 colonies are not counted.

In this lab exercise you will prepare a serial ten-fold dilution of a broth culture of *E. coli*. You will perform the pour-dish procedure to determine the number of viable colony-forming units present in 1 ml of the original broth culture.

II. DILUTING AND PLATING PROCEDURE

Materials:

4 tubes of melted nutrient agar held in a 50°C water bath

7 tubes, each containing 9 mls of sterile water

1 ml pipettes

Propipette

4 sterile Petri dishes

Cultures:

Escherichia coli

Procedure:
1. Place 7 sterile culture tubes into a culture tube rack.
2. Label the first tube 10^{-1}, and the second 10^{-2} and so on through the seventh tube (10^{-7}).
3. Use Figure 15–1 to do the serial dilution, and follow the procedure below. Also, review Fig. 15–3, which describes the use of sterile pipettes.

From *Fundamental Microbiology for the Health Care Sciences,* Fourth Edition by Frank A. Hartley, Walter Hoeksema, and Michael Ryan. Copyright © 2001 by Kendall/Hunt Publishing Company. Reprinted by permission.

4. Use a 1 ml pipet to transfer 1ml of the *E. coli* broth culture into the 9ml of sterile water in the first tube. (The **ratio** of broth culture to diluent in this tube is **1:9.** The total volume in this tube is 10ml, 1ml of which is the broth culture; the **dilution** of the broth culture is **1/10** or **10^{-1}**.)

5. Thoroughly mix the contents of the first tube.

6. Use a clean 1 ml pipet to transfer 1 ml from the first tube into the second tube. You have made a 1/10 dilution of the 1/10 dilution you made in the first tube so the final dilution in the second tube is 1/100 or 10^{-2}.

7. Thoroughly mix the contents of this tube.

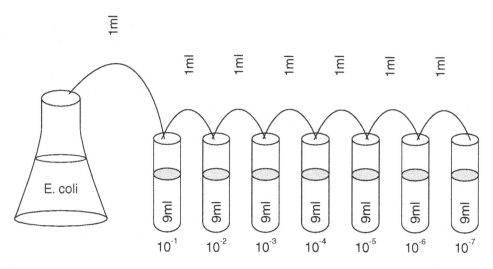

FIGURE 15–1

8. Using clean 1ml pipets for each transfer, continue the ten-fold serial dilution through the seventh tube. Be sure to thoroughly mix the contents of each tube before making a transfer from it into the next tube. The final dilution in the seventh tube will be 1/10,000,000 or 10^{-7}.

9. Label the bottoms of 4 Petri dishes, one for each of the following: 10^{-4}, 10^{-5}, 10^{-6}, and 10^{-7}. Also label each dish with your initials, label section, and the date. See **Figure 15–2.**

10. Thoroughly mix the contents of the 10^{-4} dilution tube.

11. Use a clean 1ml pipet to transfer 1ml of the content of this tube into the bottom of the Petri dish labeled 10^{-4}.

12. Remove a tube containing 16ml SABA from the water bath and wipe it dry. Gently invert the tube to mix its contents.

13. Pour the entire contents of this SABA tube into the 10^{-4} Petri dish.

14. Gently swirl the dish in a circular motion on the table top to thoroughly mix the 1ml of dilution with the SABA. Do not slosh the contents of the dish onto the lid.

15. Repeat steps 10 through 14 for each of the remaining dilutions and their respective Petri dishes.

16. Allow the SABA in each dish to solidify; invert and stack the dishes.

17. Place the stack of dishes in to a 35°C incubator for 48 hours.

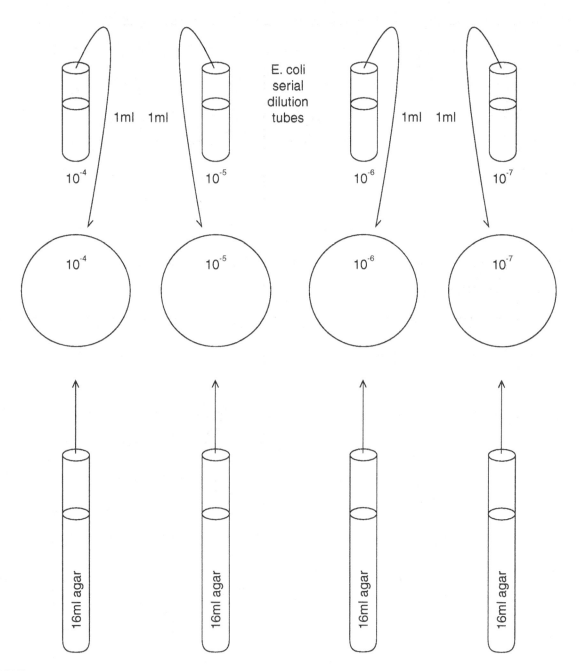

FIGURE 15–2

From *Fundamental Microbiology for the Health Care Sciences,* Fourth Edition by Frank A. Hartley, Walter Hoeksema, and Michael Ryan. Copyright © 2001 by Kendall/Hunt Publishing Company. Reprinted by permission.

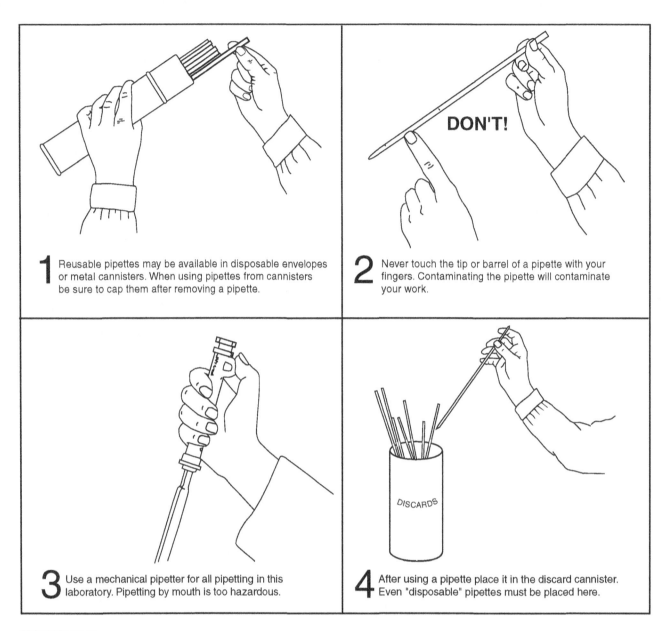

1 Reusable pipettes may be available in disposable envelopes or metal cannisters. When using pipettes from cannisters be sure to cap them after removing a pipette.

2 Never touch the tip or barrel of a pipette with your fingers. Contaminating the pipette will contaminate your work.

DON'T!

3 Use a mechanical pipetter for all pipetting in this laboratory. Pipetting by mouth is too hazardous.

4 After using a pipette place it in the discard cannister. Even "disposable" pipettes must be placed here.

DISCARDS

FIGURE 15–3

III. COUNTING AND CALCULATIONS

PROCEDURE—SECOND PERIOD

1. Remove your dishes from the incubator.
2. Examine the medium in each dish and determine which has a countable number of colonies (between 30 and 300). *E. coli* cells growing within the medium form small colonies; consider each of these as well as each typical surface.
3. Count the number of colonies on/in the medium in the dishes. Enter your data into the Worksheet.
4. Calculate the number of viable colony-forming units in 1ml of the original (undiluted) broth culture. Use the equation below.

$$\text{CFU's in 1 ml of the original broth culture} = \frac{\textbf{colonies counted} \ \times \ \textbf{dilution factor}}{\textbf{volume of dilution transferred to dish}}$$

Example: Assume you prepared a countable dish using 2ml of a 10^{-4} dilution and that you counted 76 colonies.

   a. You can express the dilution 10^{-4} as 1/10,000. The dilution factor is 10,000 or 10^{+4}.
   b. Placing the values from this example into the right side of the equation above gives you:

$$\frac{76(\text{colonies counted}) \ \times \ 10^4(\text{dilution factor})}{2(\text{volume in mls transferred to the dish})}$$

   c. When you solve the equation you obtain a value of 38×10^4 as the number of CFU's in 1 ml of the original undiluted broth culture. You must express this number in standard scientific notation — some number between 1 and 10 multiplied by some power of 10.
   d. 38, in the expression 38×10^4, is not between 1 to 10 so you must change it to 3.8 by dividing 38 by 10. Of course, 3.8×10^4 does not have the same value as 38×10^4. In order to restore the value of the expression you must multiply 10^4 by 10 to give 10^5. Thus, the value 38×10^4 expressed in standard scientific notation is 3.8×10^5.
   e. In the example then, you would report 3.8×10^5 CFU's per 1ml of original undiluted broth culture.

Name _____   Date _____

WORKSHEET

1. In the table below, record the number of *E.Coli* colonies in each of the dilution plates. If there are greater then 500 colonies, write "TNTC" ("Too numerous to count"). For plates having between 30 and 300 colonies, calculate the concentration of bacteria in the original sample, and the standard deviation.

Dilution Factor	Number of Colonies	[Bacteria] in Original Sample (CFU/ml)	Standard Deviation (±CFU/ml)
10^{-1}			
10^{-2}			
10^{-3}			
10^{-4}			
10^{-5}			
10^{-6}			
10^{-7}			
10^{-8}			

2. What do you think are the major sources of error in the serial dilution and spread plate technique?

QUESTIONS

1. List three examples of the type of information that can be provided by the viable count.

    a.

    b.

    c.

2. Define serial dilution.

3. In the space below, draw and correctly label a serial **two-fold** dilution carried out to final dilution of 1/128. Begin with a flask labeled "undiluted sample". Indicate the volume of diluents in each dilution tube, the volume transferred from tube to tube, and the final dilution in each tube.

4. If a 10^{-3} pour dish has 40 colonies, about how many colonies would you expect on 10^{-2} dish in the same series?. . . . on the 10^{-4} dish?

5. A pour dish was prepared with 3 ml of a 10^{-7} dilution of broth yeast culture. The colony count was 138. How many colony-forming units per 1ml were in the original undiluted sample? Show your work and express your answer in standard scientific notation.

Exercise 16 CULTIVATION OF ANAEROBIC BACTERIA

I. INTRODUCTION

For the cultivation of most strictly anaerobic bacteria, oxygen must be removed and the oxidation-reduction potential must be lowered. Autoclaving the media expels oxygen from the media and trace amounts of remaining oxidized molecules are scavenged up by addition of thioamino acids (reducing compounds). Usually, a reducing agent like thioglycollate is added to the medium to ensure that there is sufficient reducing power available. A small amount of agar is frequently added to broth media to help exclude oxygen. The medium usually contains an oxidation-reduction indicator dye (resazurin or methylene blue) that will be colored in the presence of oxidized molecules. The other method for the cultivation of anaerobic bacteria is the use of gas generation packages in special jars (Gas Pak @), for the production of colonies on agar plates.

Materials:
Gas Pak jars
Gas generating envelope
Thioglycollate semi-solid agar
Inoculating loops
Bunsen burners
Nutrient agar plates

Cultures:
Nutrient broth tubes or nutrient agar tubes of: *Clostridium sporogenes, Escherichia coli, and Pseudomonas aeruginosa*

II. PROCEDURE

A. *Removal of Oxygen – the Gas Pak jar*
1. Work with one jar for two bench sides. Inoculate two sets of slants with *Clostridium sporogenes, E. coli,* and *Pseudomonas aeruginosa*.
2. Put one set of slants in the Gas Pak jar, open an indicator strip and put it in the jar, open a gas-generating envelope, and 10 ml of water to the envelope and add it to the jar. Seal the jar and incubate it at 30°C until next period. Put the other set of tubes in your incubator bin at 30°C.
3. After 4 days remove the slants from the jar and observe for growth. Smell each culture through the cotton plug and correlate these smells with other experiments.

B. *Removal of oxygen by reducing agents*
1. Inoculate a tube of freshly boiled thioglycollate semi-solid agar medium with one loopful of *C. sporogenes*.

From *Microbiology: Laboratory Manual for Allied Health and General Microbiology* by Jay Sperry.

2. Inoculate a tube of nutrient broth with the same organism. Incubate both tubes at 30°C until next period.

3. Record presence or absence of growth in the two tubes.

4. Prepare a gram stain from the thioglycollate culture and observe the gram reaction and the shape of the sporangium.

5. Sketch a few cells and record the gram reaction.

III. PROCEDURE

1. Inoculate a tube of nutrient broth with a loopful of *Enterobacter aerogenes*.

2. Inoculate a second tube of nutrient broth with a loopful of *C. sporogenes*.

3. Inoculate a third tube of nutrient broth with a loopful of each organism.

4. Incubate the tubes at 37°C until next period.

5. Observe the tube the character of growth and odor.

6. Prepare gram stains from each culture showing turbidity.

7. Record your results.

From *Microbiology: Laboratory Manual for Allied Health and General Microbiology* by Jay Sperry.

IV. QUESTIONS

1. What is one theory to account for the inability of anaerobes to grow in the presence of free oxygen?

   _____

2. Why is a trace of agar put into thioglycollate media?

   _____

NOTES AND OBSERVATIONS

MICROORGANISMS AND DISEASE

Exercise 17 SELECTIVE AND DIFFERENTIAL MEDIA

INTRODUCTION

As we have seen in the previous exercises, microbes require certain nutritional components to be present in their immediate environment. Media such as tryptic soy agar and nutrient agar are complex (chemically nondefined) media that are routinely used to cultivate microorganisms. Peptones, yeast extract, and beef extract are the typical components of such complex media. These media generally permit the growth of most or all of the organisms in a mixed sample.

Sometimes a researcher only wants one particular group of microorganisms in a mixed sample to grow. In this circumstance, he or she would employ **selective media.** This type of media permits the isolation of specific groups of organisms from a mixed sample. Selective media contain compounds that inhibit the growth of one group of organisms, while allowing the growth of another group of organisms to continue. An example of such a compound is methylene blue. When methylene blue is included in media, the growth of Gram-positive organisms is inhibited, while the growth of Gram-negative organisms is permitted.

Additionally, a researcher may want to differentiate between a group of organisms that are growing in a culture. **Differential media** allow for the recognition of specific types of bacteria growing in the culture. This type of media contains substances that are utilized differently by microorganisms. The biochemical test media you used in the previous exercises are excellent examples of differential media. These media allowed you to differentiate between organisms based on their particular metabolic capabilities. For example, some bacteria were able to ferment the carbohydrate glucose, while others were not.

In some circumstances a medium can be both selective and differential. **Selective and differential media** selects for growth of a group of organisms and them differentiates between the organisms capable of growing under the selective conditions. For example, **eosin methylene blue (EMB) agar** inhibits the growth of Gram-positive bacteria due to the presence of eosin and methylene blue dyes and thus permits the growth of Gram-negative bacteria. Additionally, this medium differentiates between the Gram-negative bacteria growing on the plate by incorporating lactose in the medium. Bacteria that ferment this carbohydrate produce acids as a result of the fermentative process. The dyes in the medium register this drop in pH around the bacterial colonies and causes the colonies to take up an eosin-methylene blue dye complex. This causes the lactose positive colonies to appear blue-black. The bacteria unable to ferment this substrate do not produce acidic end products. Therefore, these colonies remain translucent.

Today's exercise will employ one type of selective medium and two types of selective and differential media. **Phenylethanol agar (PEA)** is a selective medium for the isolation of Gram-positive organisms, in particular, Gram-positive cocci, from mixed bacterial cultures. Phenylethanol is **bacteriostatic** for Gram-negative bacteria. That is, phenylethanol merely inhibits the growth of Gram-negative organisms, but does not kill them. Phenylethanol selectively inhibits DNA synthesis in Gram-negative organisms. While their growth is inhibited, it may not entirely be prevented by phenylethanol. Some Gram-negative bacteria may be present in the first quadrant, and *Escherichia coli* and *Pseudomonas aeruginosa* may occasionally grow in each quadrant.

MacConkey agar (MAC) is both a selective and differential medium. The crystal violet and bile salts are inhibitory for Gram-positive bacteria. Consequently, this medium is selective for Gram-negative organisms. This medium, like the EMB agar mentioned previously, is differential on the basis of lactose fermentation. The neutral red indication serves to contrast lactose-fermenting from lactose-nonfermenting organisms. If an organism ferments lactose, acidic by-products are released and the pH of the medium surrounding the colony drops. This drop

From *Exercises for the Microbiology Laboratory* by Elizabeth Fish McPherson. Copyright © 2001 by Elizabeth Fish McPherson. Reprinted by permission of Kendall/Hunt Publishing Company.

123

in pH causes the colony to absorb the neutral red indicator and appear pink and brick-red. Also, the drop in pH causes the precipitation of bile salts around isolated colonies. Lactose positive colonies may appear to have a hazy precipitate surrounding them. On the other hand, if the organisms do not ferment lactose, the bacterial colonies remain the color of the medium and translucent. The medium has a reddish purple coloration itself, so be sure to determine if your colonies are actually colored or simply reflect the color of the medium (translucent).

Finally, **mannitol salt agar (MSA)** is also a selective and differential medium. This medium contains 7.5% sodium chloride which results in the partial or complete inhibition of bacterial species other than halophiles (salt-loving organisms). Accordingly, MSA is selective for halophiles. This medium is differential on the basis of the halophiles to ferment the carbohydrate mannitol. The indicator phenol red is incorporated to identify mannitol-fermenting bacteria. If an organism ferments mannitol, the resulting acids lower the pH of the medium which cause the pH indicator to change from red to yellow. Therefore, colonies of bacteria that ferment mannitol with turn yellow, as will the medium immediately surrounding the colonies. Conversely, colonies of bacteria that cannot ferment mannitol will remain the color of the medium, which is pink, and will also be translucent.

Materials:

4 PEA plates
4 MAC plates
4 EMB plates
4 MSA plates
Wire loop

Cultures:

Mixed cultures of the following:
Bacillus subtilus/Proteus vulgaris
Staphylococcus aureus/Staphylococcus epidermidis
Escherichia coli/Pseudomonas aeruginosa
Enterococcus faecalis/Escherichia coli
Enterobacter aerogenes/Pseudomonas aeruginosa
Staphylococcus epidermidis/Enterobacter aerogenes
Escherichia coli/Enterobacter aerogenes

Procedure:

1. This procedure will be performed by each group, so groups should begin by carefully reviewing the procedure and by dividing the labor between group members.
2. Each group will receive four PEA plates. One plate each should be streaked for isolation with the following bacterial mixes: *Bacillus subtilis/Proteus vulgaris, Staphylococcus aureus/Staphylococcus epidermidis, Escherichia coli/Pseudomonas aeruginosa* and *Enterococcus faecalis/Escherichia coli.*
3. Each group will receive 4 MAC agar plate. One plate each should be streaked for isolation with the following bacterial mixes: *Bacillus subtilis/Proteus vulgaris, Staphylococcus aureus/Staphylococcus epidermidis, Enterococcus faecalis/Escherichia coli,* and *Enterobacter aerogenes/Pseudomonas aeruginosa.*
4. Repeat the above using 4 EMB plates.

5. Each group will receive four MSA plates. One plate each should be streaked for isolation with the following bacterial mixes: *Staphylococcus epidermidis/Enterobacter aerogenes, and Escherichia coli/Enterobacter aerogenes. Streak S. aureus and S. epidermidis onto separate plates.*

6. Label all the plates carefully.

7. The plates will be incubated for 24–48 hours at 37°C.

8. The following laboratory period, examine all the plates for growth, colony morphology, and colony coloration. Initially, groups should examine the plates for growth of the bacteria. All observations should be recorded in the worksheet. To help you in this initial examination, keep in mind that *Bacillus subtilis, Staphylococcus aureus, Staphylococcus epidermidis,* and *Enterococcus faecalis* are Gram-positive organisms, while *Proteus vulgaris, Escherichia coli, Pseudomonas aeruginosa,* and *Enterobacter aerogenes* are Gram-negatives.

9. Secondly, groups should examine the colonies on the plates for any differences in colony color. Again, all observations should be recorded in the Worksheet. Based upon this step, groups should be able to discern lactose-fermenters from non-fermenters and mannitol-fermenters from non-fermenters.

10. Based on your observations above, groups should know the identity of each organism on all agar plates.

11. Groups should perform the Gram stain procedure with each bacterium. This will result in nine Gram stains per group. Be sure to share the work and record the Gram stain reaction on the Worksheet for all organisms. When picking bacteria to make the bacterial smear, do not pick the entire colony. It is important to leave part of the colony for additional steps listed below. Also, after a Gram stain is completed, it is important to write on the agar plate what the Gram reaction and morphology are for a particular colony.

Name _____   **Date** _____

WORKSHEET

In the table below, record your results for each of the bacterial species, and their reactions to the specified media.
For the columns labeled **Growth**, record the amount of colonies as the following:

Heavy growth   +++
Medium growth   ++
Light growth   +
No growth   −

For the columns labeled **Lactose**, record the color of the colonies.
For the columns labeled **Mannitol**, record the color of the agar.

| | Phenylethanol Agar | MacConkey Agar | | EMB Agar | | Mannitol Salt Agar | | Gram Stain | |
	Growth	Growth	Lactose	Growth	Lactose	Growth	Mannitol	Stain	Morphology
B. subtilis									
P. vulgaris									
S. aureus									
S. epidermidis									
E. coli									
P. aeruginosa									
E. faecalis									
E. aerogenes									

ALTERNATE METHOD FOR DIFFERENTIAL AND SELECTIVE MEDIA

Materials:

5 Agar Plates: PEA, EMB, MAC, MSA, and TSA

Cultures:

Staphylococcus aureus

Bacillus megaterium or Bacillus subtilis

Escherichia coli

Enterobacter aerogenes

Staphylococcus epidermidis

Procedure:

1. Each group will use 5 plates and 4 bacterial organisms (2 gram positive organisms and two gram negative organisms). Example: *Staphylococcus aureus, Bacillus megaterium, E. coil, Enterobacter aerogenes* and *Staphylococcus epidermidis.*

2. Use 5 agar plates as listed below, per group and divide the plates in four using a marker on the back of each plate as follows:

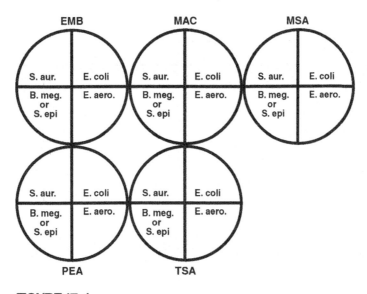

FIGURE 17–1

3. Write the name of the bacteria to be streaked in each section of the plate, as shown in Figure 17.1. Inoculate each section of the plate with that particular organism. *Note: Use a different sterile cotton applicator for each different organism. A zig-zag streak or a straight line streak can be used as indicated in Figure 17.2.

TSA

FIGURE 17–2

QUESTIONS

1. Do you see evidence that EMB and MAC agars are differential media? Selective media? Briefly explain your answers.

   _____

   _____

   _____

   _____

2. Do you see evidence that mannitol salt agar is a differential medium? A selective medium? Briefly explain your answer.

   _____

   _____

   _____

   _____

3. Describe a clinical application where you may need to use selective and differential media to separate and identify bacteria.

   _____

   _____

   _____

   _____

NOTES

Exercise 18   THE STAPHYLOCOCCI

INTRODUCTION

There are two major genera of staphylococci, *Staphylococcus* and *Micrococcus*. *Micrococcus* species are mostly nonpathogenic soil organisms and are rarely implicated in human disease. The genus *Staphylococcus* includes both pathogenic and nonpathogenic species, many of which are found among the normal flora of the skin and respiratory tract.

Staphylococci are arranged in grape-like clusters (genus *Staphylococcus*) or in tetrads (genus *Micrococcus*) as viewed under oil immersion. They are distinguished from streptococci by their colony morphology, catalase reaction, and salt tolerance. Colonies of staphylococci are usually pigmented, opaque, and are 2–3 mm in diameter after 24 hr of growth. In contrast, streptococcal colonies are usually nonpigmented, translucent, and less than 2mm in diameter. Most staphylococci are catalase-positive, but streptococci are catalase-negative. Staphylococci are tolerant to high concentrations of salt. They grow well on media containing 7.5% NaCl. On the other hand, high salt concentration generally inhibits the growth of streptococci. Mannitol salt agar is a selective medium containing 7.5% NaCl. Microbiologists use this medium to select for the growth of staphylococci while inhibiting the growth of streptococci.

The genus *Staphylococcus* contains two main species, *Staphylococcus aureus* and *Staphylococcus epidermidis*. *Staph. epidermidis* is among the normal skin flora of most persons. It is usually nonpathogenic. *Staph. aureus* also is found on the skin and mucous membranes of many individuals. Pathogenic strains of *Staph. aureus* can cause skin infections, throat infections, staphylococcal food poisoning, and toxic shock syndrome.

Staphylococcus aureus and *Staphylococcus epidermidis* are distinguished by the production of coagulase and by mannitol fermentation. *Staph. aureus* produces coagulase, an enzyme that causes blood plasma to clot or coagulate. *Staph. epidermidis* does not produce the enzyme. In addition, *Staph. aureus* ferments the sugar mannitol, but *Staph. epidermidis* does not. Mannitol fermentation can be detected on mannitol salt agar. Organisms that ferment mannitol, such as *Staph. aureus,* change the color of the agar from red to yellow. **See Fig. 18–1.**

Also, *S. aureus* cells can be **hemolytic. Hemolysin** is an exoenzyme that digests red blood cells. To test for this, a **Blood Agar** plate can be used. Blood agar contains tryptic soy agar (TSA) to which 5% sheep red blood cells have been added. If a colony of bacterial cells is produced hemolysin, there will be a round clear zone surrounding the colony because all the red blood cells have been lysed and the hemoglobin has been digested (**see Fig. 18–2**). Some strains of *Staphylococcus* are hemolytic, while others are not. Therefore, the hemolytic capability of *S. aureus* is not used as an identification characteristic of this pathogen. **See Fig. 18–2.**

In review, you can differentiate the more pathogenic *S. aureus* from *S. epidermidis,* and other Gram-positive cocci, by knowing a few specific characteristics of its enzymatic activity.

1. It produces the enzyme catalase.

2. It almost always ferments mannitol.

3. It may produce the enzyme hemolysin.

From *Basic Concepts of the Microbiology Laboratory* by Jerald D. Hendrix. Copyright © 1998 by Jerald D. Hendrix. Reprinted by permission of Kendall/Hunt Publishing Company

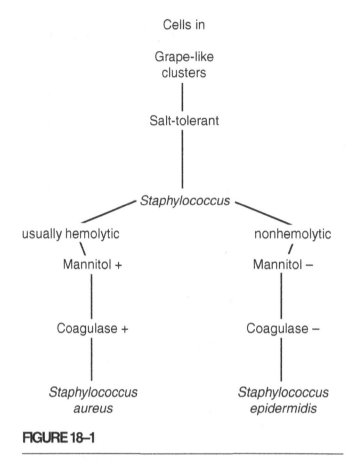

FIGURE 18–1

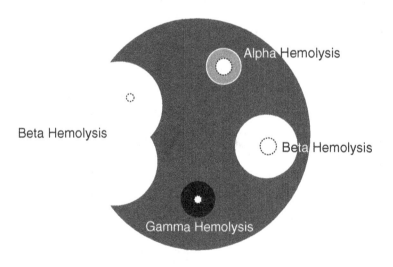

FIGURE 18–2

Beta hemolysis - **Clear zones** around the isolated colonies.
Alpha hemolysis - **Green zones** around the isolated colonies.
Gamma hemolysis - **No visible change** in the blood around the isolated colonies.

From *Microbiology: Laboratory Manual* by Jake H. Barnes and Randall M. Brand. Copyright © 1995 by Jake H. Barnes and Randall M. Brand. Reprinted by permission of Kendall/Hunt Publishing Company.

A. Blood Agar Plate for the Study of Hemolysis:

Materials:
4 Blood agar plates
Wire loop
Sterile cotton swab

Cultures:
Staphylococcus aureus
Staphylococcus epidermidis
Micrococcus luteus

Procedure:
1. Using the Quadrant streak method, streak a blood agar plate with S. *aureus*.
2. Repeat this procedure for S. *epidermidis* and M. *luteus*.
3. On the remaining blood agar plate, collect a specimen from your nose using the sterile cotton swab. Gently swab the inside membrane of your nose, then use the swab to streak the blood agar plate. Use the technique shown in Fig. 18–3.
4. Make a short, single steak with the swab on the plate. Using your wire loop, streak out the swab as shown in Fig. 18–3.

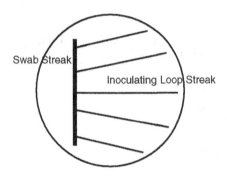

FIGURE 18–3

From *Microbiology: Laboratory Manual* by Jake H. Barnes and Randall M. Brand. Copyright © 1995 by Jake H. Barnes and Randall M. Brand. Reprinted by permission of Kendall/Hunt Publishing Company.

5. Incubate all the plates at 37°C, for 24–48 hours.
6. In the next class period, record your results on the Worksheet.
7. Also, perform a Gram stain on isolated colonies found on the plate. Record your results on the Worksheet.

B. Mannitol Salt Agar Plate, a Selective and Differential Medium:

Materials:
4 MSA plates
Wire loop
Sterile cotton Swab

Cultures:

Same as above

Procedure:

1. Using the Quadrant streak method, streak an MSA plate with *S. aureus*.
2. Repeat this procedure with *S. epidermidis* and *M. luteus*.
3. On the remaining MSA plate, streak a specimen from your nose, as described above.
4. Incubate all the plates at 37°C, for 24–48 hours.
5. In the next class period, record your results on the Worksheet.
6. Also, perform a Gram stain on isolated colonies found on the plate. Record your results on the Worksheet.

C. Slide Coagulase Test: Second Class Period

Materials:

Wire loop
Small test tubes
Nutrient broth

Cultures:

S. aureus cells from either your blood agar plate or BAP plate

Procedure:

Coagulase is an exoenzyme that catalyzes the **coagulation** of **fibrinogen** in blood plasma forming a **fibrin clot.** The formation of the fibrin clot protects the bacteria from phagocytosis. Many species of pathogenic *Staphylococcus aureus* produce coagulase, while nonpathogenic strains and *Streptococcus* species are coagulase negative.

1. Use BAP and put a loop full of bacteria into
   a. reconstituted **rabbit plasma** into a small test tube.
   b. Mix gently and **incubate** at 37.5°C for 1–2 hours.
   c. After incubation, slightly tilt and gently shake the test tube. Observe for clot formation.
   Negative – No clot is formed
   Positive – Visible clot is formed
2. Record your results on the Worksheet.

D. Catalase Test: Second class period

Materials:
Microscope slides
3% Hydrogen peroxide
Wire loop

Cultures:
S. aureus cells from either your blood agar plate or BAP plate

Procedure:
Many bacteria produce the enzyme **catalase,** which down hydrogen peroxide (H_2O_2), forming water (H_2O) and oxygen (O_2). The release of oxygen results in the formation of visible bubbles. This test is useful in distinguishing between the genera *Staphylococcus* and *Streptococcus*.
Perform the following procedure using **isolated colonies** from the blood agar plate.

1. Transfer a drop of 3% **hydrogen peroxide** (H_2O_2) to a clean microscope slide. (Place the slide on a dark surface for greater contrast.)
2. Aseptically remove a loopful of bacteria from an isolated colony on your blood **agar plate.**
3. Mix the bacteria in the drop of H_2O_2.
4. **Observations:**
   Negative – No bubbles appear.
   Positive – Bubbles in the H_2O_2.
5. After making your observations, deposit the slide in a container or disinfectant.
6. Record your results on the Worksheet.

Name _____   Date _____

WORKSHEET

A. Blood agar and MSA plates:

Bacterial Species	Hemolytic Activity	MSA medium color	Gram Stain
S. aureus			
S. epidermidis			
M. luteus			
Nose Culture			

B. Presence of Coagulase and Catalase:

Bacterial Species	Coagulase	Catalase
S. aureus		
S. epidermidis		
M. luteus		
Nose Culture		

QUESTIONS

1. List the most significant ingredient or ingredients of the following, and describe the purpose of each significant ingredient.

	Significant Ingredient	Purpose
a. Blood agar plate		
b. MSA		
c. Coagulase Test		
d. Catalase Test		

2. List three areas of the body to which staphylococci are indigenous.

   _____

3. List five pathological conditions in humans caused by S. *aureus*.

   _____

4. Why is it clinically important to be able to distinguish Staphylococcus aureus from S. epidermidis?

   _____

Exercise 19   THE STREPTOCOCCI

INTRODUCTION

The streptococcal group is a complex group containing many species. They belong to the genus *Streptococcus*. Under the oil immersion lens, most of the streptococci appear as chains of spherical cells, although certain species appear predominantly as pairs of cells. Many streptococci are normal flora in humans and other animals, where they are found in the upper respiratory tract, oral cavity, colon, and skin surfaces.

One property used to characterize the streptococci is hemolysis, the enzymatic breakdown of erythrocytes. Hemolytic reactions are determined on blood agar, a differential medium composed of tryptic soy agar to which has been added animal blood (sheep's blood is the most commonly used). Microbiologists often use blood agar as a primary isolation medium to culture bacteria from throat swabs, sputum, and other specimens. Hemolytic bacteria produce enzymes, called hemolysins, that break down the erythrocytes and cause a visible change in the blood agar.

Each bacterial species is characterized by a different hemolytic reaction, depending on the type of hemolysin that it secretes. There are three different classes of hemolytic reactions. Alpha hemolysis is a partial degradation of the erythrocytes that produces a greenish color around the bacterial colony. Bata hemolysis is the complete lysis of the erythrocytes, forming a clear zone of hemolysis surrounding the colony. Gamma reaction refers to the absence of a hemolytic reaction on blood agar. Please note that many other bacteria besides he streptococci can produce a hemolytic reaction; therefore, hemolysis by itself is not necessarily indicative of a streptococcal infection.

Most species of the genus *Streptococcus* are classified according to the Lancefield grouping system, a serological grouping system developed by Rebecca Lancefield in the 1930s. This system includes 18 groups, designated by the letters A through H and K through T. Microbiologists determine the Lancefield grouping of an unknown streptococcus by its hemolytic reaction, biochemical properties, antibiotic sensitivities, and serological reaction. Lancefield groups A, B, and D contain important human pathogens, although species in other groups also may cause infection. Two important groups of streptococci, the pneumococci (*Streptococcus pneumoniae*) and the viridans group is not included in the Lancefield grouping system because they lack the antigen that produces the differences among the Lancefield groups.

Major medical significance is in neonatal infections and in puerperal sepsis, or "childbirth fever." Different strains of this species may be alpha, beta, or gamma reactive. It is distinguished from other species of streptococci by the CAMP test. The CAMP factor is an extracellular factor that enhances the hemolytic activity of the hemolysis produced by *Staphylococcus aureus*. When the organisms are streaked at right angles to one another on a blood agar plate, the hemolysis around the group B streptococcus will look like an arrow pointing toward the *Staph. aureus* streak. This indicates the production of the CAMP factor. Streptococci other than group B do not give this reaction.

Group D includes streptococci that are also called "enterococci," because they are part of the normal flora in the large intestine of humans and other animals. The *enterococci* are commonly given the genus designation *Enterococcus* (although many authors use the genus name *Streptococcus* for these bacteria). *Enterococcus faecalis*, found in human feces, is a member of the group D streptococci. Group D streptococci are usually considered nonpathogenic, but they sometimes contribute to wound or urinary tract infections. Most members of the group are gamma reactive on blood agar, but some strains are alpha or beta hemolytic. The enterococci are distinguished from other streptococci by their ability to grow in the presence of bile and by their ability to hydrolyze the chemical esculin. One can test these properties by growing an organism on bile-esculin agar, a selective and differential medium. Enterococci will grow on this medium and will produce a brown or black color surrounding the growth; other streptococci either will fail to grow or will not produce the color change.

From *Basic Concepts of the Microbiology Laboratory* by Jerald D. Hendrix. Copyright © 1998 by Jerald D. Hendrix. Reprinted by permission of Kendall/Hunt Publishing Company.

Streptococcus pneumoniae was formerly designated *Diplococcus pneumoniae* by microbiologists. It is commonly called a "pneumococcus" because of its association with lower respiratory tract infections. *Strep. pneumoniae* is part of the normal flora of the upper respiratory tract in many individuals. It is the most frequent cause of bacterial pneumonia. It is usually alpha-hemolytic, and it may be distinguished from other streptococci by its sensitivity to optochin. Optochin disks ("O disks") are placed on blood agar plates inoculated with the bacteria. If the organism is *Strep. pneumoniae*, a zone of inhibition will be evident.

A. Blood Agar Plate for the Study of Hemolysis:

Materials:

2 Blood agar plates

Wire loop

Tongue depressor

Sterile Cotton Swab

Cultures:

Streptococcus pyogenes–demo only

Enterococcus faecalis

Streptococcus pneumoniae–demo only

Procedure:

1. Using the Quadrant streak method, streak a blood agar plate with *E. faecalis*.

2. On the remaining blood agar plate, collect a specimen from the back of your throat. If necessary, allow your lab partner to collect the specimen utilizing the tongue depressor. See Fig. 19–1 below.

3. Make a short, single streak with the swab onto the plate. Using your wire loop, streak out the swab as shown in Fig. 18–3.

4. Incubate all the plates at 37°C for 24–48 hours.

5. In the next class period, record your results on the Worksheet.

6. Also, perform a Gram stain on isolated colonies found in the plates. Record your results on the Worksheet.

FIGURE 19–1

From *Microbiology: Laboratory Manual* by Jake H. Barnes and Randall M. Brand. Copyright © 1995 by Jake H. Barnes and Randall M. Brand. Reprinted by permission of Kendall/Hunt Publishing Company.

B. Further Studies:

You can perform the test for catalase on each of the species of Streptococci. Use the same technique that was described in Exercise 18. The Streptococci are catalase negative.

Name _____   Date _____

WORKSHEET

Blood agar plates:

Bacterial Species	Hemolytic activity	Gram Stain	Cell Arrangement
S. pyogenes	β	+	
E. faecalis	γ	+	
S. pneumoniae	α	+	

QUESTIONS

1. List the disease caused by beta-hemolytic streptococci.

   _____

2. List the areas of the body where streptococci are indigenous.

   _____

3. What types of hemolysis occurred for each of the species of *Streptococcus* you tested?

   _____

4. What types of hemolysis occurred on the blood agar plate containing your throat culture?

   _____

5. Were you able to identify the type of bacteria you isolated from the throat culture?

   _____

6. The Gram stain is useful in distinguishing staphylococci from streptococci; describe the general appearance of each.

   a. staphylococci—

   b. streptococci—

Exercise 20   URINARY PATHOGENS

INTRODUCTION

The term **urogenital system** refers to organs of both the urinary system and the reproductive system of both males and females. The urinary system consists of the two kidneys, the ureters, the bladder, and the urethra. Urine passes through the kidneys via the ureters to the bladder. Upon urination, the urine is released from the bladder through the urethra. Urine and the components of the urinary system, with the exception of the portion of the urethra closest to the exterior, are sterile. Urine is very acidic and has a high salt concentration, both of which help to impede the growth of bacteria. However, as with other body systems, infections can and do occur within the urinary system. **Urinary tract infections (UTIs)** are one of the most common ailments seen in clinical medicine. This type of infection is most often caused by not completely emptying the bladder during urination. The urine remaining in the bladder becomes contaminating by bacteria from the urethra and serves as a growth medium for these bacteria. UTIs may also occur more frequently in people who do not urinate as often as they should. UTIs occur more commonly in women because their urethras are significantly shorter than men's and bacteria do not have to travel as far up the urethra to cause an infection. *Escherichia coli* is the causative agent of 80% of UTIs, but other enteric bacteria from the feces and anal region can also cause these infections. *E. coli* is also responsible for the majority of nosocomially acquired UTIs (generally through the use of catheters), but *Proteus mirabilis* is another common causative agent. These infections are diagnosed by the examination of urine for the presence of microorganisms. Typically, a clean-catch, midstream specimen is examined for the presence of a single pathogen. Even this urine sample will contain fairly substantial numbers of bacteria, but a large number of a single species is a clear indication of an infection. UTIs are easily treated with antibiotics and can be partially prevented by maintaining good personal hygiene.

The **female reproductive system** is composed of the ovaries, fallopian tubes, uterus, and vaginal canal. Most of the bacteria responsible for urinary tract infections can be placed into two groups and can be cultured using the following media.

    a. Gram-negative bacilli, such as *E. coli, Proteus, Pseudomonas, Enterobacter* and *Klebsiella,* can be cultured on EMB and MacConkey agar (see Exercise 17), or on Desoxycholate Lactose Agar (DLA). DLA contains deoxycholate and sodium citrate, which inhibit the growth of Gram-positive bacteria. The Gram-negative bacteria that grow on this medium are *E. coli* (form colonies that are red in color), and *Pseudomonas* and *Proteus* (form colonies that are white in color).

    b. Gram-positive cocci, such as *Staphylococci* and *Enterococci,* can be cultured on Blood agar plates (see Exercise 18 and 19), on MSA plates (see Exercise 18), or on Phenylethyl Alcohol Medium (PEA-B). PEA-B is inhibitory to Gram-negative organisms. If there is profuse growth, then the urinary pathogen is a Gram-positive bacterium.

Materials:
1 Blood Agar plate
1 EMB plate
1 MacConkey Agar plate
1 DLA plate
1 PEA-B plate
1 MSA plate
Wire loop

Cultures:

Simulation of urinary tract infection:

Three midstream urine samples (labeled a, b, and c), each seeded with different bacteria commonly found in urinary tract infections.

Procedure:

1. Starting with one of the seeded urine samples, streak for isolation, using the Quadrant method, on to one of each of the agar plates listed above.
2. Repeat this procedure, if your instructor assigns more than one of the urine samples to you.
3. Incubate the plates at 37°C for 24–48 hours.
4. During the next class period, examine your streak plates, and record your findings on the Worksheet.
5. Do Gram stains on identifiable colonies to corroborate your results. Enter this data onto the Worksheet.

PAYNE-APONTE ALTERNATE SIMULATION METHOD OF A URINARY TRACT INFECTION*

Two midstream simulation samples of urine are placed in a sterile plastic urine cup. The sample of urine can be used to indicate several problems such as diabetes, kidney disease and a urinary tract infection. The simulated urine sample is seeded with the following:

40 ml of TSB

2 ml of glucose

2 ml of Egg albumin

2 ml of E. coli or Proteus mirabilis

1 drop of Sheep's Blood

1 Dip stick or Chem. Strip to test the urine sample

Procedure:

1. Use the seeded urine samples and streak a TSA plate for isolation. The quadrant streak can be used.
2. Using a Chem. Strip, dip the strip into the urine sample cup and examine the Chem. Strip for color changes to detect the presence or absence of organic matter in urine. Record your results.
3. Incubate the plates at 37oC for 24-48 hours.
4. During the next class period, examine your streak plates and complete the worksheet.
5. Do Gram stains and record your results. You and your professor might continue testing the organisms by also doing antimicrobic tests.

---

*Developed by Dr. Jeanie Payne and prepared by Theresa Aponte at Bergen Community College

Name _____   Date _____

WORKSHEET

A. EMB plate:
    1. Description of colonies:
    2. Conclusions:

B. MacConkey Agar plate:
    1. Description of colonies:
    2. Conclusions:

C. DLA plate:
    1. Description of colonies:
    2. Conclusions:

D. Blood Agar plate:
    1. Description of colonies:
    2. Conclusions:

E. MSA plate:
    1. Description of colonies:
    2. Conclusions:

F. PEA-B plate:
    1. Description of colonies:
    2. Conclusions:

QUESTIONS

1. List the names and Gram reactions of seven bacteria that can be responsible for urinary tract infections.

   Name   **Gram Reaction**

   a.

   b.

   c.

   d.

   e.

   f.

   g.

2. Describe the types of bacteria found in the seeded urine samples.

   a.

   b.

   c.

3. What further testing should be done for confirmation of your results?

NOTES

Exercise 21   DENTAL CARIES SUSCEPTIBILITY

INTRODUCTION

Dental infection, including caries (cavities), rank as one of the most prevalent categories of infection worldwide. Teeth are virtually indestructible, yet, in a living human, teeth must withstand the persistent challenge of a wide array of microbes. These microbes, present in saliva and, more prominently, in plaque, produce acid which can dissolve the tooth and lead to decay. The acidogenic potential of these microbes is greatly enhanced by sucrose.

Present indications are that dental infection with *Streptococcus mutans* is associated with a significant amount of tooth decay. The presence of other acid-producing microbes in the oral cavity, such as *Lactobacillus* species, may be significant in assessing a person's propensity toward tooth decay. *S. mutans* is most frequently isolated from the surfaces of decayed teeth. *Lactobacillus* species are frequently isolated from saliva. Both of these microbes not only produce acid; they also have a high tolerance for acid accumulation in their environment. This acid-tolerance, when coupled with regular sucrose ingestion, particularly between-meal ingestion, gives these microbes a distinct advantage over other oral flora in colonizing teeth and gums.

The Snyder test was once widely used to estimate the relative number of lactobacilli in saliva. In addition, the rate at which the lactobacilli produced acid in the Snyder test medium was thought to be an indicator of the potential for caries formation, the higher the rate, the greater the caries formation potential. Although most microbial isolation and identification today involves dental plaque, the rate of acid production by the microbes present in saliva can still provide useful information about the propensity toward dental disease.

The presence of acid-producing microbes in saliva and the rate at which they produce acid is easily determined. In this lab exercise you will collect your saliva, inoculate Snyder test agar tubes with it, and monitor the rate at which acid is produced in the medium. You will use your test results to determine your propensity toward dental disease.

From *Fundamental Microbiology for the Health Care Sciences,* Fourth Edition by Frank A. Hartley, Walter Hoeksema, and Michael Ryan. Copyright © 2001 by Kendall/Hunt Publishing Company. Reprinted by permission.

Materials:

2 Melted tubes of Snyder's Test agar, held in a 45°C water bath

Sterile Petri dish

Sugarless gum

1 ml sterile pipette

Propipette

Cultures:

*Lactobacillus acidophilus from yogurt
on side bench*

Procedure:

1. Collect some saliva in the sterile Petri dish. To generate the saliva, in your mouth, your can chew on some sugarless gum.

2. Transfer 0.2ml of your saliva to a tube of melted Snyder's Test agar.

3. Roll the tube between your hand to disperse the organisms. Allow the agar to solidify.

4. Transfer 0.2ml of the *L. acidophilus* culture to the other tube of melted Snyder's Test agar. Mix well, as described above.

5. Incubate both tubes at 37°C.

6. Examine the tubes each day, for 3 days. A change of color to yellow, so that green is not dominant, is considered positive.

7. Use the table below to determine the dental caries activity. Record your results in the Worksheet.

Caries Activity

	24 hr	48 hr	72 hr
Abundant	positive		
Moderate	negative	positive	
Slight	negative	negative	positive
Negative	negative	negative	negative

Name _____   **Date** _____

WORKSHEET

The color of the medium in the uninoculated STA tube should remain blue-green throughout the 72-hour incubation.

Color of medium in uninoculated STA tube after:

24 hrs _____ 48 hrs _____ 72 hrs _____

The color of the medium in the STA tube inoculated with _Lactobacillus acidophilus_ should turn yellow after 24 hours and remain yellow for the remainder of the 72-hour incubation.

Color of the medium in STA tube with _L. acidophilus_ after:

24 hrs _____ 48 hrs _____ 72 hrs _____

STA Tube with Saliva

Time	Color change? (yes/no)	Degree of color change
24 hours		
48 hours		
72 hours		

ALTERNATE METHOD FOR DENTAL CARIES SUSCEPTIBILITY

Materials:

1 Test Tube of Snyder's Test agar per student

1 Sterile Cotton Applicator per student

Procedure:

1. Remove the sterile cotton applicator from the package and moisten it with your saliva. You will then continue to rotate the applicator as though you were brushing your teeth.

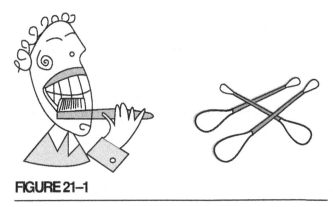

FIGURE 21–1

2. Rub the cotton applicator over the bottom row of teeth (left and right side) and the top row of teeth (left and right side) including your gums. By using this method, it will include saliva, teeth, gums and plaque.

3. After completing the procedure, remove the cap from the Snyder's Agar tube and flame the neck of the tube.

4. Place the treated cotton applicator in the Snyder's Agar tube and recap the tube.

5. Incubate the tube for 24 hours at 37°C.

6. Compare your results to the Caries Activity table.

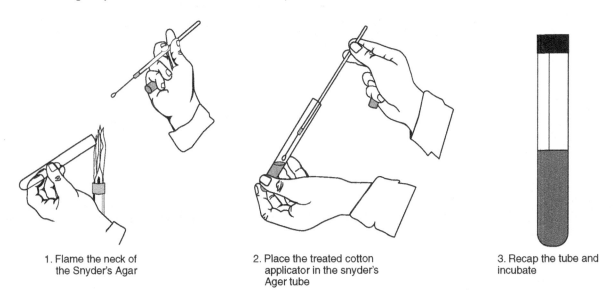

1. Flame the neck of the Snyder's Agar

2. Place the treated cotton applicator in the snyder's Ager tube

3. Recap the tube and incubate

FIGURE 21–2

QUESTIONS

1. Describe the results of your Snyder test.

   _____

   _____

   _____

   _____

2. Do your results indicate that acidogenic microbes may pose a threat to the integrity of your teeth? Explain.

   _____

   _____

   _____

   _____

3. What reasons might there be for collecting your saliva prior to brushing your teeth?. . .prior to eating?

   _____

   _____

   _____

   _____

4. If the medium in the STA tube with *L. acidophilus* had shown no color change after 72 hours, what conclusions could be drawn from any color change observed in the medium in the STA tube inoculated with your saliva?

   _____

   _____

   _____

   _____

Exercise 22   BACTERIA OF THE SKIN

INTRODUCTION

Human skin may be thought of as an ecological niche just as the bovine rumen or a hot sulfur spring are distinct niches. The skin is bathed with two secretions from glands below the surface. The eccrine and apocrine sweat glands excrete a fluid containing small amounts of nitrogenous substances and nutrients such as lactic acid. Each hair follicle is associated with a sebaceous gland which secretes sebum, a material containing lipids and fatty acids. The organic acids lower the pH of the skin surface to a pH of about 4–6, which may be inhibitory to some organisms. Fatty acids are toxic to some organisms and will play a role in determining the normal skin flora. Although the skin surface is supplied with adequate nutrients, the toxic effect of low pH and fatty acids coupled with the periodic drying of he skin makes the skin inhospitable to many organisms. It is not surprising, therefore, that the skin has a *normal flora* consisting of organisms able to survive and multiply, and a *transient flora* consisting of the organisms which are constantly coming in contact with the skin but will not remain long. The normal flora consists primarily of Gram-positive staphylococci and micrococci, *Propionibacterium acnes,* and other "diphtheroids." Other organisms which are considered normal inhabitants include the cold sore virus, certain yeasts, and the follicle mite.

In the following experiment, you will observe the bacteria from your hand before and after washing. You may find that there appears to be an increase in numbers after washing. This is due to washing away the loose scale-like *squames* forming the outermost layer of the skin and thus removing many of the transients and exposing the normal flora which is lodged around and under the squames (as well as in the hair follicles).

Materials:

2 Trypticase Soy agar plates (TSA)

Disinfectant (Zephiran, alcohol, etc.)

Soap

Hand scrub brushes

Procedure:

First Day

1. Each student should have 2 TSA plates.
2. Gently rub one index finger over one-half of the plate.
3. Scrub this finger, dry it, and then gently rub it over the other half of the petri dish. Try not to clean your other hand. You need it in the next step.
4. Immerse the other index finger in a beaker of disinfectant for about 2 minutes, dry it, and then rub it over the remaining TSA plate.
5. Incubate all plates until the following period.

Second Day

1. Determine the effect of scrubbing and immersing in disinfectant by estimating the number of colonies on the plates. Assume that both fingers had the same number of organisms to begin with.
2. Record your results in the Worksheet.

From *A Laboratory Manual for Microbiology* by J. M. Larkin. Published by Kendall/Hunt Publishing Company.

Name _____   Date _____

WORKSHEET

1. List the number of colonies found on each of the TSA plates.

2. Draw the colonies, as found on each of the plates.

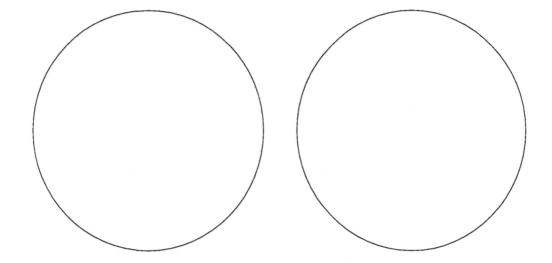

QUESTIONS

1. Describe the appearance and types of the bacterial colonies, using the colony morphology characteristics you have used earlier.

   _____

   _____

   _____

   _____

2. Compare the amount of bacteria before and after handwashing. Provide an explanation for your results.

   _____

   _____

   _____

   _____

MICROORGANISMS IN THE ENVIRONMENT

Exercise 23   UBIQUITY OF MICROORGANISMS

INTRODUCTION

Microorganisms can be found just about anyplace you look for them; from the high temperatures of hot springs to the high temperatures and pressures of undersea vents, and from the cold in your home freezer to the cold of the polar regions. This wide variety of microorganisms utilizes as wide a variety of nutritional substances. Plastic, wood, oil, toxic waste, other microorganisms, the leftover tuna casserole in the refrigerator all represent only a fraction of the nearly endless nutritional adaptations of microorganisms.

The presence of microorganisms in a specific environment usually causes no harm. There are microorganisms that occupy certain environments only transiently while others are so firmly entrenched in theirs that they are considered a normal part of it. These firmly entrenched microorganisms are often referred to as normal flora. The presence of microorganisms in some environments is necessary in maintaining a normal ecological balance. Several of the bacteria that normally live on or inside human beings help keep harmful bacteria from invading and causing disease.

In today's lab exercise, you will be employing environmental surveillance techniques, to assess both the types and numbers of microbes present in a variety of environments.

Materials:

6 Petri dishes with trypticase soy agar (TSA)

6 sterile cotton swabs

1 tube with 3 mls of sterile 0.85% NaCl solution

Wax pencil or permanent marking pen

Procedure:
A. Growing Microbes From Environmental Sources

1. Each group should obtain 6 TSA plates and a marking pencil.
2. Label the bottom of each dish for each of the following environmental sites: air, unwashed table top, table top after being disinfected, toilet seat, finger tip, and cough.
3. Remove the lid of the "air" dish and expose the agar for 30 minutes. Replace the lid after exposure.
4. Moisten a sterile cotton swab with sterile NaCl and rub the swab on the surface of your table top *before* you disinfect the area. Use the swab to streak the surface of the TSA plate. (See Fig. 23–1).
5. Moisten a sterile cotton swab with sterile NaCl and rub the swab on the surface of your table top *after* you disinfect the area. Use the swab to streak the surface of the TSA plate.
6. Moisten a sterile cotton swab with sterile NaCl and rub the swab on the surface of a toilet seat. Streak the surface of the TSA plate.
7. Rub a finger-tip on the surface of the "finger tip" TSA plate.
8. Remove the lid of the cough plate. Hold the dish close to your mouth and cough forcefully onto the surface. Replace the lid.
9. Invert all of the dishes. Incubate at 37°C for 48 hours.

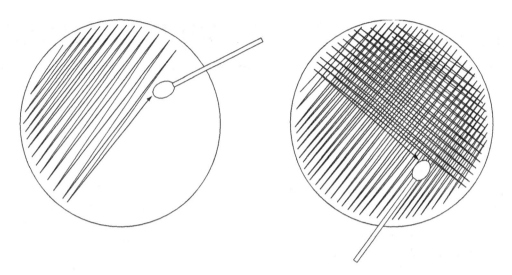

FIGURE 23–1

B. Evaluation of the Growth of Microbes From Various Environmental Sources

1. After 48 hours, record your results in Table 23–1.

2. Examine the growth on the medium in the dishes and note the number of different colony types present. If a single colony seems to cover the entire surface of the medium in a dish, examine the medium through the bottom of the dish to determine if more than a single colony is, in fact, present. Record this information on the table below.

3. Count the total number of all colonies on the medium in each dish. Record this information on the table below.

4. Answer the post-exercise questions.

TABLE 23–1

BHIA

Environmental site	# of different colonies	total # of all colonies
air		
unwashed table top		
washed table top		
toilet seat		
finger-tip		
cough		

QUESTIONS

1. From the results you obtained in this lab, what would you conclude about the use of aseptic (sterile) technique?

   _____

   _____

   _____

   _____

2. List four practical applications of environmental surveillance.

   a.

   b.

   c.

   d.

3. In general when a large variety of microbes live together in the same environment the number of individuals of any one type tends to be low. Why?

   _____

   _____

   _____

   _____

4. In general when the variety of microbes in the same environment is low the number of individuals of any one type tends to be high. Why?

   _____

   _____

   _____

   _____

5. Define normal flora.

   _____

   _____

   _____

   _____

Exercise 24   BACTERIA IN WATER

INTRODUCTION

Water is seldom, if ever, pure in nature. Water is polluted by a variety of substances, both chemical and micro-biological. Although the chemical pollutants in water are of great importance, historically most of our concern about the purity of water is related to transmission of disease by microbial pollution. Diseases such as bacillary dysentery, typhoid fever, and cholera are intestinal infections spread principally by the fecal-oral route through contaminated water sources. Therefore, detection of microorganisms present in water has become a routine pro-cedure by both scientists and water purification companies.

It would be too lengthy to identify each organism present in a contaminated water sample. Also, it would not be practical to examine the water sample for only pathogens. When such pathogens are present in water, their numbers are generally low, which would cause them to be missed during the identification process. Finally, by the time pathogens are detected, it may be too late to prevent the spread of disease caused by these organisms. It is practical, however, to examine water for the presence of only one or two organisms. The presence of these organisms indicates that pathogens may also be present. Because most diseases transmitted by water are caused by fecal organisms, scientists examine water for the presence of *Escherichia coli*, an organism that is always pres-ent in the intestinal tracts of animals and humans. Therefore, *E. coli* is an **indicator organism** or an organism whose presence in a sample suggests the presence of other, potentially pathogenic fecal organisms. *E. coli* is the most commonly used indicator organism, but any of the **coliform bacteria** can be used as indicator organisms. Coliforms are aerobic or facultatively anaerobic, Gram negative, nonsporing rods that ferment lactose with acid and gas formation within 24–48 hours. Genera of coliforms include *Enterobacter, Klebsiella, Citrobacter,* and *Escherichia*. The presence of a significant number of coliform bacteria in a water sample indicates that fecal con-tamination has occurred and that the water may not be safe for drinking or recreation.

Established public health standards specify that the maximum number of coliforms in each 100 ml sample of water depends on the intended use of the water. For example, the **standard** set for drinking water is a limit of one coliform per 100 ml of water. However, the **action limit** for drinking water, the limit at which action must be taken, is four coliforms per 100 ml of water. Water used for recreation (water sports), for irrigation, and for discharge into a bay or river have higher standard sets and action limits than drinking water.

In this exercise we will utilize two techniques to detect coliforms in water samples. The first is the **mem-brane filter method,** which traps the microbes on filter paper. The second technique, the **multiple-tube fermentation method,** utilizes a three-stage test which narrows the identification to fecal coliforms or eliminates them altogether.

I. MEMBRANE FILTER METHOD

The membrane filter technique uses a membrane filter with pore sizes around 0.45 micrometers. Bacteria are filtered out of the water sample and trapped on the membrane filter. The membrane filter is then transferred onto an absorbent pad saturated with a differential media such as m-Endo Broth. The plate is incubated for 24 hours at 35 degrees C and organisms that have been trapped on the filter will grow colonies which can be counted. A characteristic metallic sheen will appear from the coliform colonies.

Materials:

100 ml water sample

Membrane filter apparatus

Absorbent pad

Sterile forceps

m-Endo broth

50mm Petri dish

Sterile pipettes

Procedure:

Lab Day One

1. Obtain a 100 ml water sample for testing.
2. Assemble a membrane filtering device for use. See Figure 24–1 below.
3. Place a membrane filter (pore size 0.45 micrometers) on the filtration unit with the grid side up.
4. Carefully pour the 100 ml sample into the filter funnel on the top of the filtering apparatus.

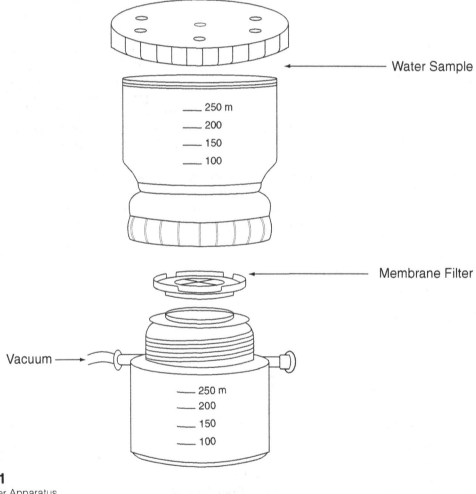

FIGURE 24–1

Membrane Filter Apparatus

5. Connect the vacuum to the filtering apparatus and allow the water to filter through the membrane. Any bacteria will be trapped on the membrane filter.

6. Place a sterile absorbent pad in a sterile 50 mm Petri dish and add 2 ml of fresh Endo broth to saturate the pad.

7. With sterile forceps transfer the membrane filter, grid side up, to the pad in the Petri dish.

8. Incubate the plate at 35 degrees C for 22–24 hours. Invert the Petri plate in the incubator.

Lab Day Two

9. Remove the plate from the incubator, remove the membrane filter from the dish and place on absorbent paper for 30 minutes.

10. Count the total number of colonies and those that have a green, metallic sheen luster. Express these colony counts per 100 ml of water on the work sheet for the Membrane Filter Method.

II. MULTIPLE TUBE FERMENTATION METHOD

The multiple tube fermentation method is based upon three stages of testing. The first stage inoculates the water sample to lactose broth. The tubes are incubated and observed for gas production. If gas production occurs, then one can presume that coliforms are possible in the sample. If gas is not observed, then coliforms are presumed not to be present. See Fig. 24–2.

If the presumptive test is positive for gas, then the confirmed test is second in our series. Gram-positive organisms can also form gas from lactose, therefore this test confirms the presence of coliform bacteria by using a selective and differential media such as eosin methylene blue (EMB) and endo broth. EMB contains lactose and methylene blue which is inhibitory for Gram-positive bacteria. Endo broth contains basic fuchsin, which causes coliform bacteria to form light-pink to rose colored colonies.

The bacteriological analysis of water is completed in the third stage of testing if coliforms are confirmed. In the completed test, colonies of coliforms from the confirmed test are once again incubated in lactose broth, but at a higher temperature of 45 degrees C for 24 hours. Coliform bacteria can tolerate this higher temperature and will produce acid and gas, others will not. Finally a Gram stain is performed upon the coliforms from the confirmed test. If Gram-negative rods are observed, then the test is completed, positively confirmed coliform bacteria in the original sample of water.

Fecal coliforms can be identified from the non-fecal coliforms using the IMViC tests.

Materials:
Day One –
Water sample
Sterile pipettes
2 Durham tubes of double strength lactose broth (DSLB)
2 Durham tubes of single strength lactose broth (SSLB)

Further testing –
EMB Agar plates
Nutrient agar slant
2 Durham tubes of SSLB
Gram stain material
IMViC materials

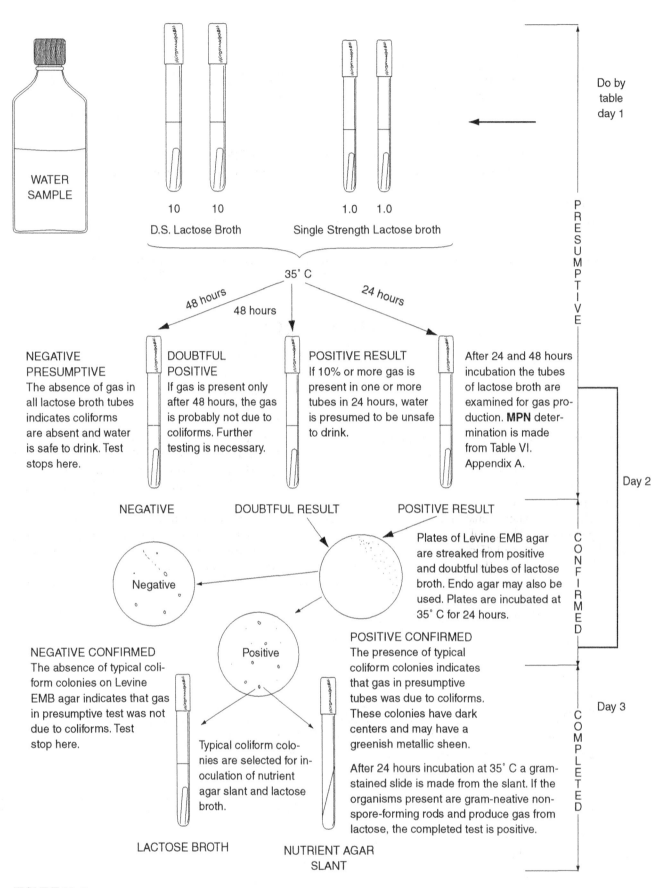

WATER SAMPLE

10   10

D.S. Lactose Broth

1.0   1.0

Single Strength Lactose broth

35° C

48 hours

48 hours

24 hours

NEGATIVE PRESUMPTIVE

The absence of gas in all lactose broth tubes indicates coliforms are absent and water is safe to drink. Test stops here.

DOUBTFUL POSITIVE

If gas is present only after 48 hours, the gas is probably not due to coliforms. Further testing is necessary.

POSITIVE RESULT

If 10% or more gas is present in one or more tubes in 24 hours, water is presumed to be unsafe to drink.

After 24 and 48 hours incubation the tubes of lactose broth are examined for gas production. **MPN** determination is made from Table VI. Appendix A.

NEGATIVE

DOUBTFUL RESULT

POSITIVE RESULT

Negative

Positive

Plates of Levine EMB agar are streaked from positive and doubtful tubes of lactose broth. Endo agar may also be used. Plates are incubated at 35° C for 24 hours.

NEGATIVE CONFIRMED

The absence of typical coliform colonies on Levine EMB agar indicates that gas in presumptive test was not due to coliforms. Test stop here.

Typical coliform colonies are selected for inoculation of nutrient agar slant and lactose broth.

POSITIVE CONFIRMED

The presence of typical coliform colonies indicates that gas in presumptive tubes was due to coliforms. These colonies have dark centers and may have a greenish metallic sheen.

After 24 hours incubation at 35° C a gram-stained slide is made from the slant. If the organisms present are gram-neative non-spore-forming rods and produce gas from lactose, the completed test is positive.

LACTOSE BROTH

NUTRIENT AGAR SLANT

Do by table day 1

PRESUMPTIVE

CONFIRMED

COMPLETED

Day 2

Day 3

FIGURE 24-2

Based on figure from *Microbiological Applications* by Harold J. Benson. 1998 McGraw-Hill Companies, Inc.

A. Presumptive Test

1. Each student will inoculate two 10 ml tubes of double-strength lactose broth with 10 ml of sewage sample A or B, (but not both) and two tubes of single-strength lactose broth with 1 ml of the same sewage sample. (Note: To satisfy statistical requirements, in actual water analysis more than 2 tubes of each medium would be inoculated.) Incubate the 4 tubes at 37°C for 24 hrs.

2. After 24 hours, observe for fermentation of lactose to acid and gas. Acid production is indicated by a yellow color, and gas production is indicated by a bubble in the inverted Durham tube. Durham tubes should have 10% (or more) of their volume displaced with gas, before gas production is considered positive. If less than 10% gas volume is observed, reincubate the tubes at 37°C an additional 24 hours.

3. If the proper amount of gas is observed, proceed to the confirmed test. If no or insufficient gas is observed in 48 hrs., record a negative result and assume that no coliforms (and no enterics and/or fecal contamination) are present.

4. Record your data on the Worksheet.

B. Confirmed Test

1. If gas is produced within the lactose broth, streak a loopful of the culture on an EMB plate, and Endoagar plate.

2. Incubate for 48 hours, at 37°C.

3. Observe plates for characteristic metallic sheen from coliform colonies. Any pigmented colonies are considered to be Gram-negative colonies due to the EMB agar inhibiting the growth of Gram-positive bacteria.

4. Record your data on the Worksheet.

Adapted from *Laboratory Manual for General Microbiology* by H.E. Urban. Published by Kendall/Hunt Publishing Company.

C. Completed Test

1. Transfer some growth from the identified colonies on the EMB plate to a lactose broth tube and to a nutrient agar slant.

2. Incubate nutrient agar slant for 24 hours at 35 degrees C.

3. Incubate the lactose broth at 45 degrees C for 24 hours.

4. Prepare a Gram stain of the growth from the nutrient slant. Identify coliforms as Gram negative rods.

5. Observe the lactose broth for the production of a gas and acid conditions. Acid and gas production are indicative of coliforms.

6. Record your data on the Worksheet.

D. IMViC Test

1. If coliforms were identified by the completed test, then the final step would include identification of *E. coli* from *E. aerogenes*. The series of tests that would accomplish this is IMViC. See **Exercises 29 and 31** for a description of this test.

Name _____   Date _____

WORKSHEET

Membrane Filter Data Table

Number of Colonies With Metallic Green Luster	
Total Number of Colonies	

Multiple Tube Fermentation Method Data Tables

Presumptive Test

Gas Production	

Key

+ = Gas Production

− = No Gas Production

Confirmed Test

Green Metallic Colonies or Purple Colonies	

Key

+ = Colonies Present

− = Colonies Not Present

Completed Test

Acid Production	
Gas Production	

Key

A = Acid (yellow)

− = No Acid (pink-red)

G = Gas Production

− = No Gas Production

IMViC Test

Indole	Methyl-Red	V-P	Citrate

Key
+ = Positive Test
− = Negative Test

QUESTIONS

1. Why are coliforms used as indicators in the bacteriological examination of water?

   _____

   _____

   _____

2. Does the presence of feces in water mean that pathogens are present?

   _____

   _____

   _____

3. What are advantages and disadvantages of the membrane filter method?

   _____

   _____

   _____

4. Why are organisms other than coliform bacteria not sought in the bacteriological examination of water?

   _____

   _____

   _____

5. Indicate several water borne diseases and their etiological agents.

   _____

   _____

   _____

6. List several conditions and circumstances under which the bacteriological examination of water might be important.

   _____

   _____

   _____

III. ALTERNATE MULTIPLE TUBE FERMENTATION METHOD

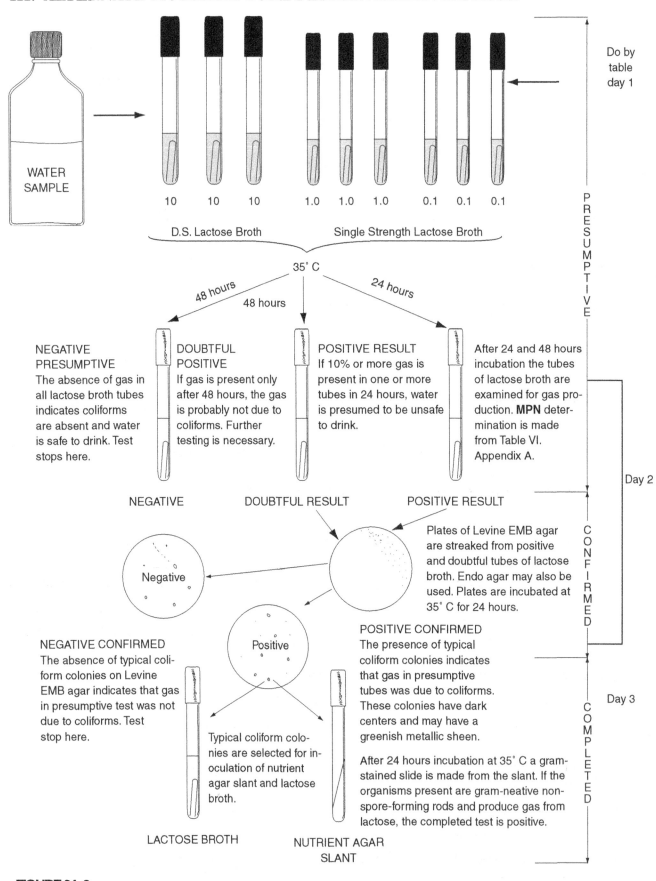

FIGURE 24–3

Based on figure from *Microbiological Applications* by Harold J. Benson. 1998 McGraw-Hill Companies Inc.

NOTES

Exercise 25   BACTERIA IN MILK

INTRODUCTION

Louis Pasteur discovered that mild heating of beer and wine helped prevent spoilage, but did not seriously alter the taste of these products. This process later became known as **pasteurization** and is now principally associated with milk. Modern milking and milk handling have significantly reduced the number of microbes present in raw milk, but contamination can occur at several points. Cow udders and teats are covered with bacteria and these microorganisms easily contaminate milk during the actual milking process, although milk itself is normally a sterile liquid. Hand drawn milk will naturally contain more microorganisms than mechanically drawn milk because the milk is exposed to the atmosphere for a longer period of time. Storing, transporting, and processing the milk also present situations where contamination can occur. The most commonly isolated organisms in milk are *Staphylococcus epidermidis, Micrococcus sp., Escherichia coli, Staphylococcus aureus,* and *Salmonella sp.* The sources of these organisms are cow udders and teats, the bodies of the cows, and unsanitary handling of milk.

Milk was first pasteurized to eliminate the bacterium *Mycobacterium tuberculosis,* which is the causative agent of tuberculosis. The pasteurization temperature was raised in 1956 to ensure that the organism *Coxiella burnetii,* the causative agent of Q fever, was killed. This disease is characterized by fever, chills, headache, malaise, sweats, and other evidence of a pneumonia-type infection. Endocarditis occurs in 5–10% of cases and is invariably fatal. *C. burnetii* is an obligately parasitic, intracellular bacterium that is transmitted among cattle through tick bites. The organisms are shed in the feces, milk, and urine of infected cattle.

Prior to the discovery of the presence of *C. burnetii* in milk, pasteurization was accomplished by heating milk to **63°C for 30 minutes.** Today, milk is pasteurized by heating to **72°C for at least 15 seconds.** This process is known as the **flash method** of pasteurization, while the former is known as the **holding method** of pasteurization. The flash method, in addition to ridding the milk of *C. burnetii,* also reduces the total bacterial count of the milk, which allows milk to keep longer under refrigeration. Many relatively heat-resistant microorganisms survive these two pasteurization processes, but these organisms are unlikely to cause disease or cause the milk to spoil in a short period of time. It is important to realize that pasteurized milk does still contain bacteria which include *Streptococcus lactis* and species of *Lactobacillus.* It is these bacteria that cause milk to spoil even when stored in the refrigerator. When these organisms produce enough lactic acid to cause the pH of the milk to fall below 4.8, the proteins in the milk begin to coagulate. Coagulated or soured milk is still safe to drink, but the taste and the appearance of the milk generally cause it to be undesirable for use. Sterile milk is also available in Europe and, to a limited extent, in the United States. Sterile milk has a "cooked" flavor but can be kept under room temperature conditions as long as the container remains sealed. This is useful in areas of the world where refrigeration facilities are not readily available. This process is also used on some of he containers of coffee creamers found in restaurants in the United States. The process by which milk is sterilized is known as the **ultrahigh temperature (UHT) processing** procedure. The temperature of milk is raised from 74°C to 140°C and then dropped back to 74°C in less than five seconds.

As with water and food, routine plate counts are performed on samples of milk by food processing companies and public health agencies. The presence of a large number of bacteria is undesirable in milk because they increase the likelihood that pathogens will be present in the milk and that the milk will spoil. Again, only those bacteria capable of growing under the environmental conditions provided will grow and these may only represent a small number of the total number of bacteria found in the milk.

I. STANDARD PLATE COUNT

One of the most reliable indicators of the sanitary quantity of the milk is the number of viable bacteria present per milliliter. U.S. Public Health Services have set standards for the numbers according to a **Standard Plate Count** procedure. The numbers obtained include a total number of coliforms and a total number of microorganisms per milliliter of milk (See **Table 25–1** below). In this exercise, you will analyze a milk sample utilizing a Standard Plate Count.

TABLE 25–1
Bacterial Count Table

Milk Type	Total Bacteria Count/ml	Coliforms/ml
Raw milk	$\leq 100,000$	≤ 150
Pasteurized milk	$\leq 20,000$	≤ 5

Materials:

5 Sterile pipettes

5 Melted TGEA pours, in the 60 C water bath

5 Sterile 9ml dilution blanks of water

Milk samples

5 Petri dishes

Procedure:

1. Pipet 1ml of the milk sample into a 9 ml dilution blank. Mix the milk and the sterile water by rotating the tube thoroughly between the palms. This is your 10^{-1} sample.
2. Pipet 1 ml of the 10^{-1} sample into another 9 ml dilution blank. Mix thoroughly between the palms. This represents your 10^{-2} sample.
3. Pipet 1 ml of the 10^{-2} sample into another 9 ml dilution blank. Mix thoroughly between the palms. This represents your 10^{-3} sample.
4. Pipet 1 ml of the 10^{-3} sample into another 9 ml dilution blank. Mix thoroughly between the palms. This represents your 10^{-4} sample.
5. Pipet 1 ml of the 10^{-4} sample into another 9 ml dilution blank. Mix thoroughly between the palms. This represents your 10^{-5} sample. See **Figure 25–1** below.
6. Pipet 1 ml into bottom of plate and once they are all in the plates then pour in agar on top of sample. Gently swirl agar to cover plate.
7. Repeat procedure 6 for all your dilutions.
8. Allow to solidify and incubate at 37 degrees C for 48 hours.
9. During the next lab period, obtain incubated plates.
10. Choose the plate which has between 25–300 colonies. Count and record on the Worksheet. Calculate your total bacterial count.

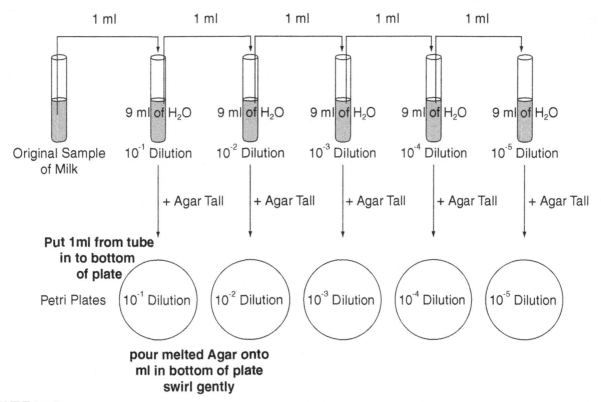

1 ml   1 ml   1 ml   1 ml   1 ml

9 ml of H₂O   9 ml of H₂O   9 ml of H₂O   9 ml of H₂O   9 ml of H₂O

Original Sample 10^{-1} Dilution 10^{-2} Dilution 10^{-3} Dilution 10^{-4} Dilution 10^{-5} Dilution
of Milk

+ Agar Tall   + Agar Tall   + Agar Tall   + Agar Tall   + Agar Tall

**Put 1ml from tube
in to bottom
of plate**

Petri Plates 10^{-1} Dilution 10^{-2} Dilution 10^{-3} Dilution 10^{-4} Dilution 10^{-5} Dilution

**pour melted Agar onto
ml in bottom of plate
swirl gently**

FIGURE 25–1
Serial Dilutions

From *Microbiology: Laboratory Manual* by Jake H. Barnes and Randall M. Brand. Copyright © 1995 by Jake H. Barnes and Randall M. Brand.
Reprinted by permission of Kendall/Hunt Publishing Company.

Name _____   Date _____

WORKSHEET

Dilution	Number of Colonies
10^{-1}	
10^{-2}	
10^{-3}	
10^{-4}	
10^{-5}	

Multiply the number of bacterial colonies by the dilution factor for the appropriate plate counted with 25–300 colonies. Record this as the total bacterial count per ml of milk.

Total Number of Bacteria per ml of milk equals:

$$\frac{\underline{\hspace{3cm}}}{\text{(colony count)}} \times \frac{\underline{\hspace{3cm}}}{\text{(dilution factor)}}$$

$$\frac{\underline{\hspace{5cm}}}{\text{(TBC)}}$$

II. SELECTIVE MEDIA FOR COUNTING COLIFORMS

This exercise uses a selective medium, violet red bile agar (VRB), to determine the coliform count in 1 ml of contaminated milk. On this medium, coliforms ferment the sugar lactose, forming red subsurface colonies surrounded by a reddish zone of precipitated bile. Other organisms, such as Gram-positive bacteria, are inhibited by the bile salts and crystal violet.

Materials:

A sample of milk containing some coliform bacteria

Tubes containing 20 ml of melted violet red bile agar in a 45°C water bath

Sterile empty petri dishes

1 ml pipettes

Procedure:

1. Working in pairs, pipette 1.0 ml of the milk sample into a sterile petri dish.
2. Obtain a tube of melted agar from the water bath and pour it promptly into the petri dish containing the 1.0 ml of milk.

3. Keeping the dish flat on the bench, mix the milk with the agar by moving the plate in a figure eight pattern. Work quickly to complete mixing before the agar hardens but smoothly as not to slop the melted agar onto the lid. Let the plant stand until the agar is hard.

4. At the option of the instructor, an agar overlay step may be added. In this case, obtain a tube of overlay agar and pour it promptly on top of the hardened violet red bile agar. Do not swirl the plate but tilt it carefully to make the overlay run evenly over the surface.

5. When the agar is hard, invert the plate and incubate at 37°C for 48 hours.

9. Examine the plates for red subsurface colonies which are surrounded by a zone of precipitated bile.

10. Count the coliform colonies and record your findings: Number of coliforms/ml of milk _____

QUESTIONS

1. Why is it helpful to know the actual number of coliforms in milk?

_____

_____

_____

_____

2. What is the purpose of crystal violet and bile salts in the violet red bile agar plates?

_____

_____

_____

_____

3. At what temperature does agar melt?

_____

_____

_____

_____

4. Are coliforms aerobic, anaerobic or facultative? What does this have to do with using an overlay agar?

_____

_____

_____

_____

NOTES

Exercise 26   BACTERIA IN FOOD

INTRODUCTION

Microbial diseases of the intestinal tract are the second leading cause of illness in the United States. Only respiratory diseases cause more illnesses per year. Most of the diseases of the gastrointestinal tract are caused by people ingesting food or water contaminated with microorganisms or their toxins. We discussed contamination of water in the previous exercise, so now we will discuss contamination of food and food products.

Gastrointestinal diseases are causes by two distinct mechanisms. In a **foodborne infection** a person ingests food contaminated by microorganisms. Signs and symptoms of such an infection are somewhat delayed in their appearance because the bacteria must first multiply in the gut and invade the intestinal tissues. Symptoms of a foodborne infection include diarrhea, cramps, nausea, vomiting, and fever, and are typically seen 8–12 hours following ingestion of the contaminated food. Occasionally, the bacteria can spread throughout the entire body from the intestinal mucosa, as is the case with typhoid fever. Another name for a foodborne infection is **bacterial enteritis.** Salmonellosis, typhoid fever, shigellosis, cholera, and traveler's diarrhea are examples of diseases caused by this type of mechanism. Another category of gastrointestinal disease is caused not by the contaminating organisms themselves, but by the toxins produced by the contaminating organisms. In a **microbial intoxication** microorganisms grow and multiply in a food source and produce toxins that remain in the food. The ingestion of these preformed toxins produces the illness. Signs and symptoms are caused by the toxins, soothe onset of symptoms is generally more rapid than in a foodborne infection. Signs and symptoms of diarrhea, cramps, nausea, and vomiting appear within a few minutes to a few hours. Fever is typically not a symptom of a microbial intoxication. Examples of microbial intoxications are staphylococcal enterotoxicosis, botulism, and mycotoxicoses, which are intoxications caused by fungal toxins.

Microbial contamination of food and food products is largely a product of industrialization. Large food processing plants provide ideal environments for cross contamination of large quantities of food with microorganisms. The primary source of bacteria is the intestinal tracts of processed animals. Business and institutions that serve food to large numbers of people also present increased risk for illness. A batch of food that becomes contaminated is served to many people, all of who may become sick. On the contrary, an individual who prepared food for himself/herself only risks making one person sick.

As with water testing, food is examined for the presence of **coliforms,** which are indicative of fecal contamination. Recall that *E. coli* is the most commonly used indicator organism and is a Gram negative, facultatively anaerobic, nonsporing rod that ferments lactose. Typically, food and food products are examined for the presence of these organisms and an upper limit of 50–100 coliforms per gram of food is deemed acceptable. However, if the number of coliforms per gram of food exceeds 100, the food is judged to be unsafe. One problem with testing food in this manner is that testing takes 24 to 48 hours and by that time the food may have been distributed. Recent food recalls in the United States have highlighted this problem. Other methods for ensuring the safety of food are being investigated. It is important to note, however, that the absence of coliforms does not mean that the food is safe to eat. Contamination by pathogenic organisms not found in fecal material can and does occur.

News media frequently report on outbreaks of **salmonellosis,** which is a foodborne infection caused by members of the genus *Salmonella.* Estimates are that there are over two million cases of this disease every year, although most cases are unreported. Contamination of chicken with *Salmonella* occurs periodically as does contamination of eggs and egg products. Evidence has shown that bacteria are transmitted to the eggs before the eggs are laid. If these eggs are eaten raw, as in ice cream, or partially cooked, as in eggs fried "sunny side up," the bacteria are ingested and can cause disease. Prevention of microbial contamination of food and the resulting illnesses in humans can be accomplished by good sanitation practices and by cooking food fully. Cutting contaminated meat on a cutting board that is then used to cut raw vegetables allows microorganisms present in the meat to spread to the vegetables. Although the organisms in the meat will be killed during cooking, the organisms remain alive on the raw vegetables. Contaminating of cutlery also occurs by this manner. Additionally, bacteria can be transferred to the hands of the chef. The chef can then contaminate any other food products he/she prepares if proper precautions (washing the hands) are not taken. A simple, easy, and cheap method of ridding the kitchen environment of microorganisms is by cleaning utensils, cutting boards, and counter surfaces with a dilute solution of bleach. Poultry and other animals serve as reservoirs for *Salmonella*, so these organisms will continue to persist in food, but carefully practices in the kitchen can help prevent further spread of these bacteria and the illnesses they cause.

I. BACTERIAL CONTAMINATION OF HAMBURGER

Materials:

Hamburger meat

4 Petri dishes

One 99ml sterile water blank

4 melted agar pours, in a 60 degree C water bath

Sterile pipettes

Sterile water

Procedure:

1. Label 4 petri plates with your group member, type of food and the dilution factor for each plate. The dilution factor for each plate will be: 10^{-1}, 10^{-2}, 10^{-3}, 10^{-4}.
2. Weigh 10 grams of the food sample to be tested. Blend the 10 grams of food with the 90 ml sterile water sample. See **Figure 26–1** below, for the completion of steps 2–7. (If a blender is not available, shake the mixture for 5–10 minutes.)
3. Pipet 1 ml of the blended food and water sample mixture to your 10^{-1} plate. This represents a 1:10 dilution.
4. Pipet 0.1 ml of the mixture of the 10^{-2} plate. This is a 1:100 dilution plate.
5. Now pipet 1 ml of the sample to the 99 ml water sample. Vigorously shake the mixture.
6. Pipet 1 ml of the mixture prepared in Step 5 to the 10^{-3} plate. This is your 1:1000 dilution.
7. Pipet 0.1 ml of the mixture prepared in step 5 to the 10^{-4} plate. This is the 1:10000 dilution.
8. Pour a liquefied agar tall into each of the 4 petri plates. To ensure a mixture of the sample and the agar, gently swirl each of the plates.
9. Allow the plates to solidify and then incubate for 48 hours at 37 degrees C.
10. During the next lab period, obtain incubated plates.
11. Choose the plate which has between 25–300 colonies. Count and record on the data table.

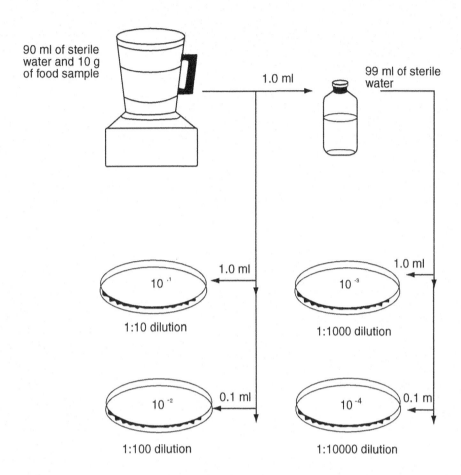

90 ml of sterile
water and 10 g
of food sample

1.0 ml

99 ml of sterile
water

1.0 ml

10^{-1}

1:10 dilution

1.0 ml

10^{-3}

1:1000 dilution

0.1 ml

10^{-2}

1:100 dilution

0.1 m

10^{-4}

1:10000 dilution

FIGURE 26–1
Standard Plate Count of Food

From *Microbiology: Laboratory Manual* by Jake H. Barnes and Randall M. Brand. Copyright © 1995 by Jake H. Barnes and Randall M. Brand. Reprinted by permission of Kendall/Hunt Publishing Company.

Name _____   Date _____

WORKSHEET

Dilution	Number of Colonies
10^{-1}	
10^{-2}	
10^{-3}	
10^{-4}	

    Multiply the number of bacterial colonies by the dilution factor for the appropriate plate counted with 25–300 colonies. Record this as the total bacterial count per ml of the food sample.

Total Number of Bacteria per ml of food equals:

$$\underline{\hspace{3cm}} \times \underline{\hspace{3cm}}$$
(colony count)   (dilution factor)

$$\underline{\hspace{4cm}}$$
(TBC)

From *Microbiology: Laboratory Manual* by Jake H. Barnes and Randall M. Brand. Copyright © 1995 by Jake H. Barnes and Randall M. Brand. Reprinted by permission of Kendall/Hunt Publishing Company.

II. BACTERIAL CONTAMINATION OF CHICKEN

We will determine the number of viable bacteria in a sample of chicken skin using the **heterotrophic plate count** technique. In this technique a medium which supports the growth of most heterotrophic organisms (those requiring organic carbon) is utilized. Dilutions of the blended chicken skin will be placed in an empty petri plate, tempered plate count agar poured into the plate, the plate swirled to evenly distribute any microorganisms, and the agar allowed to solidify. Following incubation, plates containing between 25 and 250 organisms will be counted. This number is divided by the dilution of the chicken skin plated and divided by the grams of chicken skin in the original sample as is shown in the formula below:

Number of bacteria per gram of chicken skin = cfu counted × dilution plated ÷ wt. of sample

This results in the number of bacteria per gram of chicken skin. The larger the count, the greater the chance that pathogenic organisms are present. This method has a major limitation, however. Only those organisms capable of growing under the conditions provided will grow. The actual number of bacteria per gram may be much larger.

From *Exercises for the Microbiology Laboratory* by Elizabeth Fish McPherson. Copyright © 2001 by Elizabeth Fish McPherson. Reprinted by permission of Kendall/Hunt Publishing Company.

Procedure:

1. This procedure will be performed by each group, so groups should begin by carefully reviewing the procedure and by dividing the labor between groups members. Review Figure 26–2 prior to beginning this experiment.

2. Each group will receive five empty, sterile petri plates.

3. The plates should be labeled as follows:

   10^{-2}

   10^{-3}

   10^{-4}

   10^{-5}

   10^{-6}

   Each label should also contain the information always included on the labels.

4. Each group will receive two 99 ml dilution blanks. One should be labeled 10^{-3} and the other should be labeled 10^{-5}.

5. Each group will receive a chicken skin sample at 10^{-1}. Some settling may have occurred, so the sample should be vortexed prior to using.

6. Using a sterile 1 ml pipet, transfer 1 ml of the chicken skin sample to the bottle labeled 10^{-3} and shake the bottle vigorously 30 times.

7. Again, using a sterile 1 ml pipet, transfer 1 ml of the 10^{-3} chicken skin sample to the bottle labeled 10^{-5} and shake the bottle vigorously 30 times.

8. Using sterile 1 ml pipets and aseptic technique, pipet the following amounts of each dilution into appropriately labeled plates.

   0.1 ml of the 10^{-1} dilution to the plate labeled 10^{-2}.

   1 ml of the 10^{-3} dilution to the plate labeled 10^{-3}.

   0.1 ml of the 10^{-3} dilution to the plate labeled 10^{-4}.

   1 ml of the 10^{-5} dilution to the plate labeled 10^{-5}.

   0.1 ml of the 10^{-5} dilution to the plate labeled 10^{-6}.

9. Remove one plate count agar pour from the 45°C water bath and pour the contents into one of the plates. Rotate the plate gently to ensure mixture of the chicken skin dilution with the agar. Repeat this procedure with the remaining four agar pours.

10. Allow the agar plates to solidify completely.

11. Rinse the tubes that contained the agar using tap water and place these tubes in the area your teaching assistant has designated.

12. The plates will be incubated at 37°C for 24 hours.

13. The following laboratory period, count the cfu's on the plates and calculate the total number of microorganisms per gram of chicken skin. Record your results the Worksheet.

14. Dispose of the plates by placing them in the biohazard bag located in your laboratory.

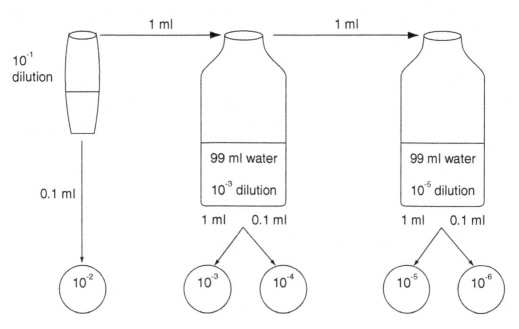

FIGURE 26–2
Schematic of how to aliquot volumes of diluted chicken skin

Name _____   Date _____

WORKSHEET

Petri Plate	Dilution	ml Plated	Final Dilution	No. Colonies	CFU/ml
1	_____	_____	_____	_____	_____
2	_____	_____	_____	_____	_____
3	_____	_____	_____	_____	_____
4	_____	_____	_____	_____	_____
5	_____	_____	_____	_____	_____

Total number of microorganisms per gram of chicken skin:

Exercise 27 BACTERIA IN SOIL

INTRODUCTION

Most people normally associate bacteria with disease or food spoilage and therefore believe these microorganisms are to be avoided. It is important to recognize that disease-causing bacteria make up only a small fraction of the total number of microorganisms and that microorganisms are essential for life. Life on earth would not exist in its present form if it were not for the activities of microorganisms. For example, these organisms are responsible for the degradation of organic matter which replenishes the soil with nutrients such as carbon dioxide, nitrates, nitrites, and nitrogen gas. Therefore, bacteria play a role in the cycling of elements on the earth. The soil is one of the main reservoirs of microorganisms on the earth. Soil contains bacteria, fungi, protozoa, algae, and viruses, each specialized to survive in its own particular niche. The most numerous of all microbes in the soil are the **bacteria.** A gram of normal garden soil contains millions of bacteria. **Actinomycetes,** the second most common microorganism found in the soil, are known as filamentous bacteria, but are usually considered separate from bacteria. These organisms are the source of antibiotics such as streptomycin and tetracycline. **Fungi,** the third most populous microorganism in the soil, are also important members of the microbial community.

The numbers and varieties of microorganisms found in the soil indicate that species are capable of surviving under many different conditions. Soil varies greatly with respect to its physical features, such as oxygen, light, moisture, pH, temperature, and nutrients, and therefore, the microorganisms present will vary. For example, acidic soils will contain a larger number of fungal species, compared to alkaline soils. Soil also contains obligate aerobes, facultative anaerobes, and obligate anaerobes because it has a wide range of oxygen conditions depending on the depth and the water levels of the local water table. As the usable nutrients and suitable environmental conditions, such as light, temperature, and oxygen, change, so do the populations of microorganisms.

From *Exercises for the Microbiology Laboratory* by Elizabeth Fish McPherson. Copyright © 2001 by Elizabeth Fish McPherson. Reprinted by permission of Kendall/Hunt Publishing Company.

Materials:
Soil suspension (1 g in 500 ml)
2 tubes with 15–20 ml melted nutrient agar at 45° C
1 tube with 15–20 ml melted Sabouraud agar at 45° C
45° C water bath
Sterile 1.0 ml pipette
3 sterile petri dishes

Procedure:
1. Pipette 1.0 ml of soil suspension into each of 2 petri dishes, and 0.1 ml into the third dish.
2. Add nutrient agar to one dish containing 1.0 ml of soil suspension and to the dish with 0.1 ml.
3. Add Sabouraud agar to the remaining dish with 1.0 ml of soil suspension.
4. Swirl the dishes to disperse the organisms.
5. After the agar has solidified, incubate the plate at room temperature for about 5 days.
6. During the next lab period, examine each plate for the presence of colonies surrounded by zones of inhibition. These may be small and are often more easily seen by holding the plate up to the light.

From *A Laboratory Manual for Microbiology* by J. M. Larkin. Published by Kendall/Hunt Publishing Company.

Name _____   **Date** _____

DATA SHEET

Sabourand Agar	TSA
Number of colonies	Number of colonies
1 ml	1 ml
0.1 ml	

From *A Laboratory Manual for Microbiology* by J. M. Larkin. Published by Kendall/Hunt Publishing Company.

IDENTIFICATION OF UNKNOWN BACTERIA

One of the most interesting experiences in introductory microbiology is to attempt to identify an unknown microorganism that has been assigned to you as a laboratory problem. The next four exercises pertain to this phase of microbiological work. You will be given one or more cultures of bacteria to identify. The only information that might be given to you about your unknowns will pertain to their sources and habitats. All the information needed for identification will have to be acquired by you through independent study.

The first step in the identification procedure is to accumulate information that pertains to the organism's morphological, cultural and physiological (biochemical) characteristics. This involves making different kinds of slides for cellular studies and the inoculation of various types of media to note the growth characteristics and types of enzymes produced. As this information is accumulated, it is recorded in an orderly manner on the Worksheet, which is located at the end of Exercise 31.

After sufficient information has been recorded, the next step is to consult a taxonomic key, which enable one to identify the organism. For this final step *Bergey's Manual of Systematic Bacteriology* will be used. Copies of volumes 1 and 2 of this book will be available in the laboratory, library, or both. In addition, a CD-ROM computer simulation program called *Identibacter interactus* may be available, which can be used for identifying and reporting your unknown.

Success in this endeavor will require meticulous techniques, intelligent interpretation, and careful record-keeping. Your mastery of aseptic methods in the handling of cultures and the performance of inoculations will show up clearly in your results. Contamination of your cultures with unwanted organisms will yield false results, making identification hazardous speculation. If you have reason to doubt the validity of the results of a specific test, repeat it; *don't rely on chance!* As soon as you have made an observation or completed a test, record the information on the Worksheet. Do not trust your memory – record data immediately!

Exercise 28 MORPHOLOGICAL STUDY OF UNKNOWN

INTRODUCTION

The first step in the identification of an unknown bacterial organism is to learn as much as possible about its morphological characteristics. One needs to know whether the organism is rod-, coccus-, or spiral-shaped; whether or not it is pleomorphic; its reaction to gram staining; and the presence or absence of endospores, capsules, or granules. All this morphological information provides a starting point in the categorization of an unknown.

Figure 28–1 illustrates the steps that will be followed in determining morphological characteristics of your unknown. Note that fresh broth and slant cultures will be needed to make the various slides and perform motility tests. Note that gram staining, motility testing, and measurements will be made from the broth culture; gram staining and other stained slides will also be made from the agar slant. The rationale as to the choice of broth or agar slants will be explained as each technique is performed.

As soon as morphological information is acquired be sure to record your observations on the Worksheet at the back of the manual. Proceed as follows:

Materials:

gram-staining kit

spore-staining kit

acid-fast staining kit

Loeffler's methylene blue stain

nigrosine or India ink

tubes of nutrient broth and nutrient agar

1 nutrient gelatin deep

TSA plates

New Inoculations

For all of these staining techniques you will need 24–48 hour cultures of your unknown. If your working stock slant is a fresh culture, use it. If you don't have a fresh broth culture of your unknown inoculate a tube of nutrient broth and incubate it at its estimated optimum temperature for 24 hours.

Gram's Stain

Since a good gram-stained slide will provide you with more valuable information than any other slide, this is the place to start. Make gram-stained slides from both the broth and agar slants, and compare them under oil immersion.

Two questions must be answered at this time: (1) Is the organism gram-positive, or is it gram-negative? and (2) Is the organism rod- or coccus- shaped? If your staining technique is correct, you should have no problems with the Gram reaction. If the organism is a long rod, the morphology question is easily settled; however, if your organism is a very short rod, you may incorrectly decide it is coccus-shaped.

Keep in mind that short rods with round ends (coccobacilli) look like cocci. If you have what seems to be a coccobacillus, examine many cells before you make a final decision. Also, keep in mind that *while rod-shaped organisms frequently appear as cocci under certain growth conditions, cocci rarely appear as rods. (Streptococcus mutans* is unique in forming rods under certain conditions.) Thus, it is generally safe to assume that if you have a slide on which you see both coccuslike cells and short rods, the organism is probably rod-shaped. This assumption is valid, however, only if you are not working with a contaminated culture!

Record the shape of he organism and its reaction to the stain on the Worksheet.

Cell Size

Once you have a good gram-stained slide, determine the size of the organism with an ocular micrometer. If the size is variable, determine the size range. Record this information on the Worksheet.

Motility and Cellular Arrangement

If your organism is a nonpathogen make a wet mount or hanging drop slide from the broth culture. This will enable you to determine whether the organism is motile, and it will allow you to confirm the cellular arrangement. By making this slide from broth instead of the agar slant, the cells will be well dispersed in natural clumps. Note whether the cells occur singly, in pairs, masses, or chains. *Remember to place the slide preparation in a beaker of disinfectant when finished with it.*

If your organism happens to be a pathogen do not make a slide preparation of the organisms; instead, stab the organism into a tube of semisolid or SIM medium to determine motility. Incubate for 48 hours.

Be sure to record your observations on the Worksheet.

Endospores

If your unknown is a gram-positive rod, check for endospores. *Only rarely is a coccus or gram-negative rod a spore-former.* Examination of your gram-stained slide made from the agar slant should provide a clue, since endospores show up as transparent holes in gram-stained spore-formers. Endospores can also be seen on unstained organisms if studied with phase-contrast optics.

If there seems to be evidence that the organism is a spore-former, make a slide using the spore-staining technique. *Since some spore-formers require at least a week's time of incubation before forming spores, it is prudent to double-check for spores in older cultures.*

Record on the Worksheet whether the spore is terminal, subterminal, or the middle of the rod.

Other Structures

If the protoplast in gram-stained slides stains unevenly, you might wish to do a simple stain with Loeffler's methylene blue for evidence of metachromatic granules.

Also, a capsule stain may be performed at this time.

Gelatin Stab Culture

Make a stab inoculation into the gelatin deep by stabbing an inoculating needle (straight wire) into the medium to the bottom of the tube. Pull it straight out.

After the culture has been properly incubated, place the tube in an ice water bath for 30 minutes.

Remove your tube of nutrient gelatin from the ice water bath and examine it. Check first to see if liquefaction has occurred. Organisms that are able to liquefy gelatin produce the enzyme *gelatinase*.

Record your data on the Worksheet.

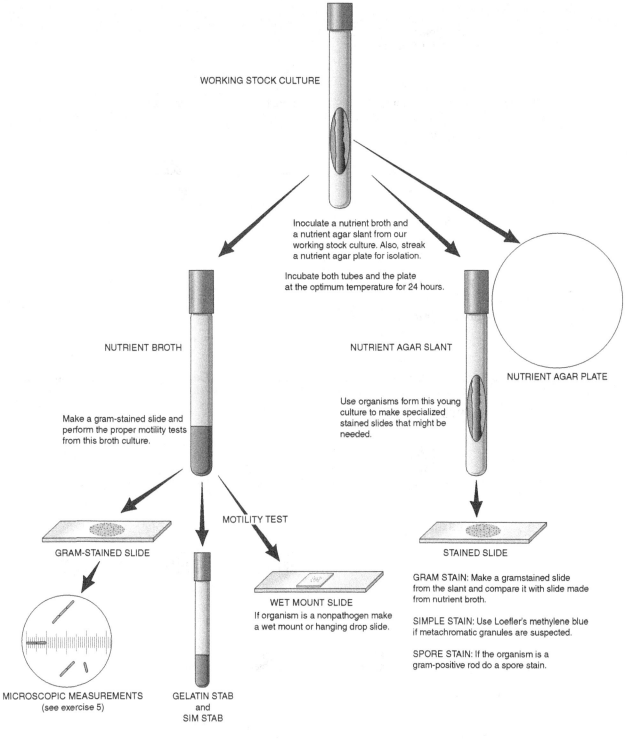

WORKING STOCK CULTURE

Inoculate a nutrient broth and
a nutrient agar slant from our
working stock culture. Also, streak
a nutrient agar plate for isolation.

Incubate both tubes and the plate
at the optimum temperature for 24 hours.

NUTRIENT BROTH

NUTRIENT AGAR SLANT

NUTRIENT AGAR PLATE

Make a gram-stained slide and
perform the proper motility tests
from this broth culture.

Use organisms form this young
culture to make specialized
stained slides that might be
needed.

MOTILITY TEST

GRAM-STAINED SLIDE

STAINED SLIDE

WET MOUNT SLIDE

If organism is a nonpathogen make
a wet mount or hanging drop slide.

GRAM STAIN: Make a gramstained slide
from the slant and compare it with slide made
from nutrient broth.

SIMPLE STAIN: Use Loefler's methylene blue
if metachromatic granules are suspected.

SPORE STAIN: If the organism is a
gram-positive rod do a spore stain.

MICROSCOPIC MEASUREMENTS
(see exercise 5)

GELATIN STAB
and
SIM STAB

FIGURE 28-1
Procedure for morphological study

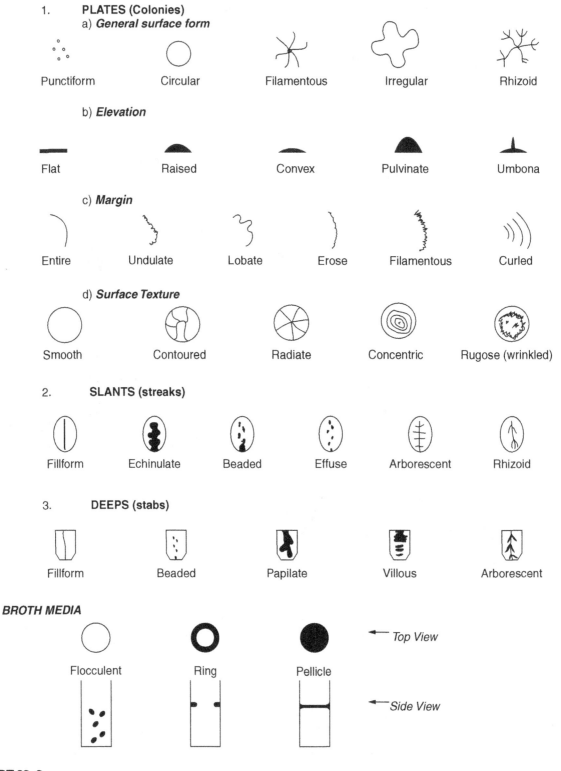

FIGURE 28–2
Colony characteristics

WORKSHEET FOR UNKNOWN BACTERIA

1. Choose an unknown. Write your name on the back of the TSA plate or TSA (slant) tube.

2. Do the following with your unknown: Transfer it onto (2) TSA plates and designate the temperatures 25°C and 37°C. Use the continuous streaking method or the quadrant streaking method.

3. Inoculate a *Nutrient gel* tube and a *SIM* tube.

4. Inoculate (1)TSB tube (Broth Tube).

*Using your stock culture, do the following: a. Smear preparation—Exercise 7, b. The Gram Stain- Exercise 9, c. Motility Test—The Wet Mount Slide—Suspend a small amount of the microorganism in sterile water and add a cover glass.

*Complete the index card with the following information and hand it in to your professor:

Your Name _____Unknown Number _____

Gram Reaction _____

Morphology (shape) _____

Motility: Motile or Non-motile _____

NOTES

Exercise 29 PHYSIOLOGICAL CHARACTERISTICS: BIOOXIDATIONS

INTRODUCTION

Although morphological and cultural characteristics are essential in getting to the genus, species determination requires a good deal more information. The physiological information that will be accumulated here and in the next two exercises will make species identification possible.

Since all physiological (biochemical) reactions in organisms are catalyzed by enzymes, and since each enzyme is produced by individual genes, we are, essentially, formulating a genetic profile of an organism as we discover what enzymes are produced. The physiological characteristics of concern in this exercise pertain to the chemistry of biooxidations. The enzymatic reactions that fall in this category pertain to respiration and fermentation.

Biooxidation reactions in bacteria pertain to the manner in which they get their energy: some are oxidative and others are fermentative. Strict aerobes that oxidize organic substances to produce the end products carbon dioxide and water are said to be **oxidative.** By utilizing organic compounds as electron donors, with oxygen as the ultimate electron (and hydrogen) acceptor, they produce CO_2 and water as end products to release energy. The ability to utilize free oxygen is accomplished by a cytochrome enzyme system. This process is also called **respiration.**

Fermentative bacteria are organisms that also utilize organic compounds for energy but lack a cytochrome system. Instead of producing only CO_2 and H_2O, they produce complex end products, such as acids, aldehydes, and alcohols, that are oxidizable and reducible. In these organisms oxygen is not the ultimate electron acceptor, and the reactions occur under anaerobic conditions. Various gases, such as carbon dioxide, hydrogen, and methane, are also produced. In fermentative bacteria the organic compounds act both as electron donors and electron acceptors.

Sugars, particularly glucose, are the compounds most widely used by fermenting organisms. Other substances such as organic acids, amino acids, purines, and pyrimidines also can be fermented by some bacteria. The end products of the particular fermentation are determined by the nature of the organism, the characteristics of the substrate, and environmental conditions such as temperature and pH.

Although fermentation and respiration represent two different types of energy-yielding biooxidations, they can both be present in the same organism, as is true of facultative anaerobes. In the presence of molecular oxygen these organism shift from fermentation to respiration. An exception, however, is seen in the lactic acid bacteria where fermentation occurs in the presence of air (O_2).

Six types of biooxidation reactions will be studied in this exercise: (1) Durham tube sugar fermentations, (2) mixed acid fermentation, (3) butanediol fermentation, (4) catalase production, (5) oxidase production, and (6) nitrate reduction.

The performance of all these test on your unknown will involve a considerable number of inoculations because a set of positive test controls will also be needed. Although photographs of positive test results are provided in this exercise, seeing the actual test results in test tubes will make it more meaningful.

As you perform these various tests, attempt to keep in mind what groups of bacteria relate to each test. Although some tests are not very specific in pointing the way to unknown identification, others are very narrow in application.

One last comment of importance: *It is not routine practice to perform all these tests in identifying every unknown.* Although it might appear that our prime concern here is to identify an organism, our most important goal is to learn about the various types of test for biooxidation enzymes that are available. The use of unknown bacteria to learn about them simply makes it more of a challenge. In actual practice physiological tests are used very selectively. The "shotgun approach" employed here is used to expose you to the multitude of tests that are available.

I. FIRST PERIOD (INOCULATIONS)

The following two sets of inoculations (unknown and test controls) may be done separately or combined into one operation. The media for each set of inoculations are listed separately under each heading.

Unknown Inoculations

Figure 29–1 illustrates the procedure for inoculation seven test tubes and one Petri plate with your unknown. Since your instructor may want you to inoculate some different sugar broths, blanks have been provided in the materials list for write-ins. *If different media are distinguished from each other with differently colored tube caps, write down the colors after each medium below.*

Materials:
for each unknown:
Durham tubes with phenol red indicator
1 glucose broth
1 lactose broth
1 mannitol broth
1 sucrose broth
2 MR-VP medium
1 nitrate broth
1 nutrient agar slant
1 Petri plate of trypticase soy agar (TSA)

1. Label each tube with the number of your unknown and an identifying letter as designed in figure 29–1.
2. Label one half of the Petri plate UNKNOWN and the other half *P. AERUGINOSA*.
3. Inoculate all broths and the slant with a loop. Inoculate one half of the TSA plate with your unknown, using an isolation technique.

Test Control Inoculations

Figure 29–3 illustrates the procedure that will be used for inoculating five test tubes to be used for positive test controls. The Petri plate shown on the right side is the same one that is shown in figure 29–1; thus, it will not be listed in the materials list.

Materials:
1 glucose broth (Durham tube)
2 MR-VP medium
1 nitrate broth
1 nutrient agar slant
nutrient broth cultures of *Escherichia coli, Enterobacter aerogenes, Staphylococcus aureus, and Pseudomonas aeruginosa*

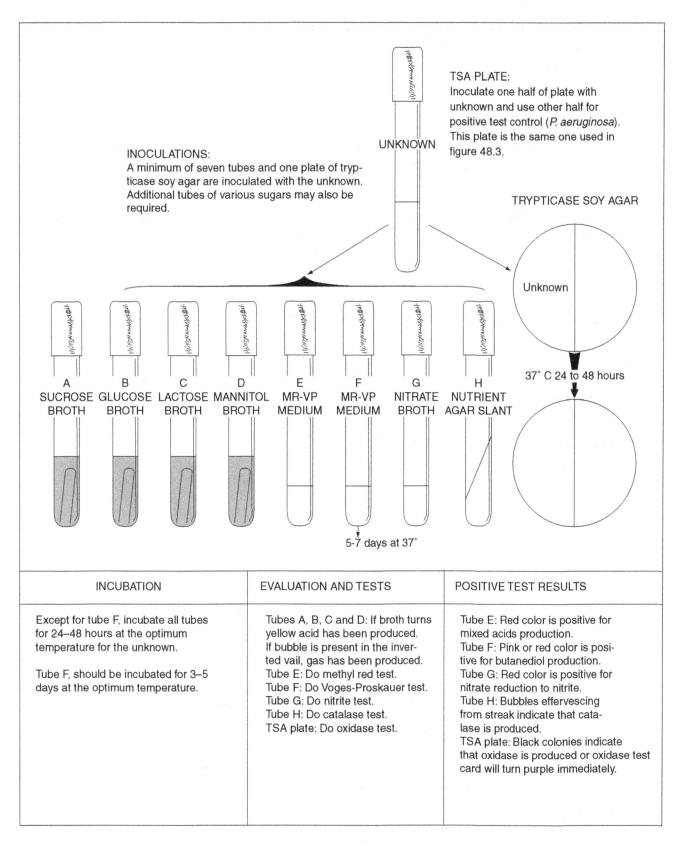

INOCULATIONS:
A minimum of seven tubes and one plate of trypticase soy agar are inoculated with the unknown. Additional tubes of various sugars may also be required.

UNKNOWN

TSA PLATE:
Inoculate one half of plate with unknown and use other half for positive test control (*P. aeruginosa*). This plate is the same one used in figure 48.3.

TRYPTICASE SOY AGAR

Unknown

37° C 24 to 48 hours

A	B	C	D	E	F	G	H
SUCROSE BROTH	GLUCOSE BROTH	LACTOSE BROTH	MANNITOL BROTH	MR-VP MEDIUM	MR-VP MEDIUM	NITRATE BROTH	NUTRIENT AGAR SLANT

5-7 days at 37°

INCUBATION	EVALUATION AND TESTS	POSITIVE TEST RESULTS
Except for tube F, incubate all tubes for 24–48 hours at the optimum temperature for the unknown. Tube F, should be incubated for 3–5 days at the optimum temperature.	Tubes A, B, C and D: If broth turns yellow acid has been produced. If bubble is present in the inverted vail, gas has been produced. Tube E: Do methyl red test. Tube F: Do Voges-Proskauer test. Tube G: Do nitrite test. Tube H: Do catalase test. TSA plate: Do oxidase test.	Tube E: Red color is positive for mixed acids production. Tube F: Pink or red color is positive for butanediol production. Tube G: Red color is positive for nitrate reduction to nitrite. Tube H: Bubbles effervescing from streak indicate that catalase is produced. TSA plate: Black colonies indicate that oxidase is produced or oxidase test card will turn purple immediately.

FIGURE 29–1
Procedure for unknown biooxidation tests

1. Label each tube with the code letter assigned to it as listed:

   glucose broth   A[1]
   MR-VP medium   E[1]
   MR-VP medium   F[1]
   nitrate broth   G[1]
   nutrient agar slant H[1]

2. Inoculate each of these tubes with a loopful of the appropriate test organism according to figure 29.3.

3. Inoculate the other half of the TSA plate with *P. aeruginosa*.

Incubation

Except for tube F (MR-VP), all the unknown inoculations should be incubated for 24-48 hours at the unknown's optimum temperature. Tube E should be incubated for 3-5 days at the optimum temperature.

Except for Tube F[1] of the test controls, incubate all the test control tubes and the TSA plate at 37° C for 24-48 hours. Tube F[1] should be incubated at 37° C for 3-5 days.

II. SECOND PERIOD (TEST EVALUATIONS)

After 24 to 48 hours incubation, arrange all your tubes (except tubes F and F[1]) in a test-tube rack in alphabetical order, with the unknown tubes in one row and the test controls in another row. As you interrupt the results, record the information on the Worksheet immediately. Don't trust your memory. Any result that is not properly recorded will have to be repeated.

Durham Tube Sugar Fermentations

When we use a bank of Durham tubes containing various sugars, we are able to determine what sugars an organism is able to ferment. If an organism is able of ferment a particular sugar, acid will be produced and gas *may* be produced. The presence of acid is detectable with the color change of a pH indicator in the medium. Gas production is revealed by the formation of a void in the inverted vial of the Durham tube. If it were important to know the composition of the gas, we would have to use a Smith tube as shown in figure 29–2. For our purposes here the Durham tube is preferable.

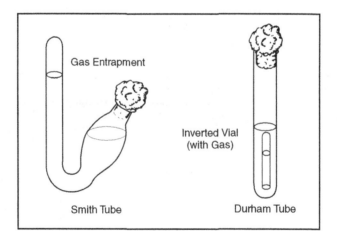

FIGURE 29–2
Two types of fermentation tubes

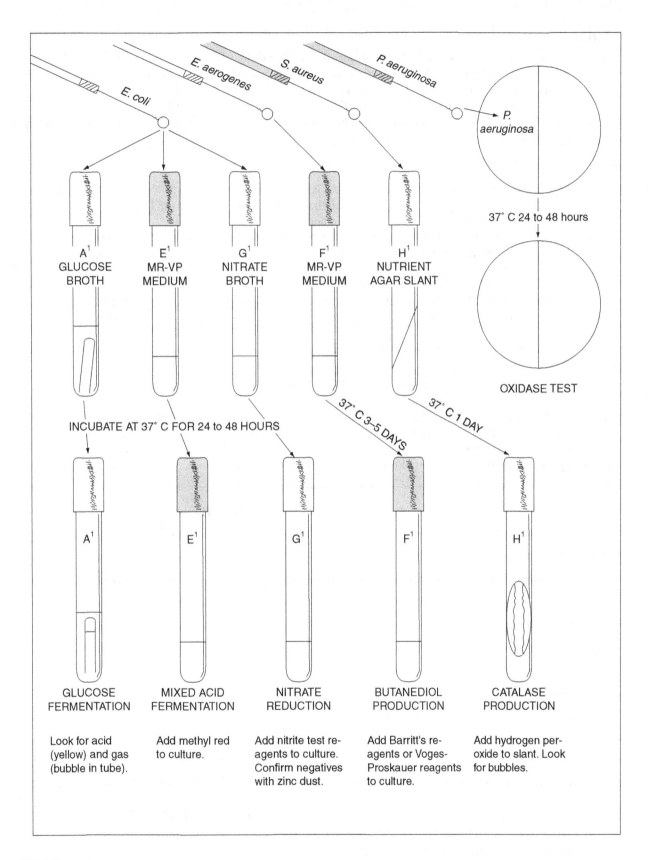

FIGURE 29-3

Test control procedure for biooxidations

Media

The sugar broths used here contain 0.5% of the specific carbohydrate plus sufficient amounts of beef extract and peptone to satisfy the nitrogen and mineral needs of most bacteria. The pH indicator phenol red is included for acid detection. This indicator is red when the pH is above 7 and yellow below this point.

Although there are many sugars that one might use, glucose, lactose, mannitol and sucrose are logical ones to begin with. Your instructor may have had you included one or more additional kinds, and it is very likely that you may wish to use some others later.

Interpretation

Examine the glucose test control tube (tube A^1) that you inoculated with *E. coli*. Note that the phenol red has turned yellow, indicating acid production. Also, note that the inverted vial has a gas bubble in it. These observations tell us that *E. coli* ferments glucose to produce acid and gas.

Now examine the three sugar broths (tubes A, B, C, and D) that were inoculated with your unknown and record your observations on the Worksheet. If there is no color change, record NONE after the specific sugar. If the tube is yellow with no gas, record ACID. If the inverted vial contains gas and the tube is yellow, record ACID AND GAS.

An important point to keep in mind at this time is that *a negative result on an unknown is as important as a positive result.* Don't feel that you have failed in your technique if many of your tubes are negative!

Mixed Acid Fermentation (Methyl Red Test)

A considerable number of gram-negative intestinal bacteria can be differentiated on the basis of the end products produced when they ferment the glucose in MR-VP medium. Genera of bacteria such as *Escherichia, Salmonella, Proteus,* and *Aeromonas* ferment glucose to produce large amounts of lactic, acetic, succinic, and formic acids, plus CO_2, H_2, and ethanol. The accumulation of these acids lowers the pH of the medium of 5.0 and less.

If methyl red is added to such a culture, the indicator turns red, an indication that the organism is a *mixed acid fermenter.* These organisms are generally great gas producers, too, because they produce the enzyme *formic hydrogenylase,* which splits formic acid into equal parts of CO_2 and H_2.

$$HCOOH \xrightarrow{\text{formic hydrogenylase}} CO_2 + H_2$$

Medium

MR-VP medium is essentially a glucose broth with some buffered peptone and dipotassium phosphate.

Procedure:

Perform the methyl red test first on your test control tube (E^1) and then on your unknown (tube E). Proceed as follows:

Materials:

dropping bottle of methyl red indicator

1. Add three or four drops of methyl red to test control tube E^1, which was inoculated with *E. coli*. The tube should become red immediately.

   A **reddish color,** as shown in the left-hand tube of the middle illustration of figure 28–4 is a positive methyl red test.

2. Repeat the same procedure with your unknown culture (tube E) of MR-VP medium. If your unknown culture becomes yellow, your unknown is negative for this test.

3. Record your results on the Worksheet.

Butanediol Fermentation (Voges-Proskauer Test)

A negative methyl red test may indicate that the organism being tested produced a lot of 2,3 butanediol and ethanol instead of acids. All species of *Enterobacter* and *Serratia*, as well as some species of *Erwinia, Bacillus*, and *Aeromonas*, do just that. The production of these non-acid end products results in less lowering of the pH in MR-VP medium, causing the methyl red test to be negative.

Unfortunately, there is no satisfactory test for 2,3 butanediol; however, acetoin (acetylmethylcarbinol), a precursor of 2,3 butanediol, is easily detected with Barritt's reagent.

Barritt's reagent consists of alpha-naphthol and KOH. When added to a 3- to 5-day culture of MR-VP medium, and allowed to stand for some time, the medium changes to pink or red in the presence of acetoin. Since acetoin and 2,3 butanediol are always simultaneously present, the test is valid. This indirect method of testing for 2,3 butanediol is called the *Voges-Proskauer* test.

Procedure:

Perform the Voges-Proskauer test on your unknown and test control tubes of MR-VP medium (tubes F and F^1). Note that the test control tube was inoculated with *E. aerogenes*. Follow this procedure:

Materials:

Barritt's reagents

2 pipettes (1 ml size)

2 empty test tubes

1. Label one empty test tube **F** (for unknown) and the other **F^1** (for control).
2. Pipette 1 ml from culture tube **F** to the empty tube **F** and 1 ml from culture tube **F^1** to the empty tube **F^1**. Use separate pipettes for each tube.
3. Add 18 drops (about 0.5 ml) of Barritt's solution A (alpha-naphthol) to each of these tubes that contain 1 ml of culture.
4. Add an equal amount of Barritt's solution B (KOH) to the same tubes.
5. Shake the tubes vigorously every 20 seconds until the control tube (F^1) turns pink or red. Let the tubes stand for one or two hours to see if the unknown turns red. *Vigorous shaking is very important* to achieve complete aeration. A positive Voges-Proskauer reaction is **pink** or **red.**
6. Record your results on the Worksheet.

Catalase Production

Most aerobes and facultatives that utilize oxygen produce hydrogen peroxide, which is toxic to their own enzyme systems. Their survival in the presence of this antimetabolite is possible because they produce an enzyme called *catalase*, which converts the hydrogen peroxide to water and oxygen:

$$2H_2O_2 \xrightarrow{\text{catalase}} 2H_2O + O_2$$

It has been postulated that the death of strict anaerobes in the presence of oxygen may be due to the suicidal act of H_2O_2 production in the absence of catalase production. The presence or absence of catalase production is an important means of differentiation between certain groups of bacteria.

Procedure:

To determine whether or not catalase is produced, all that is necessary is to place a few drops of 3% hydrogen peroxide on the organisms of a slant culture. If the hydrogen peroxide effervesces, the organism is catalase-positive.

Materials:

3% hydrogen peroxide

test control tube H[1] with *S. aureus* growth and unknown tube H

1. While holding test control tube H[1] at an angle, allow a few drops of H_2O_2 to flow slowly down over the *S. aureus* growth on the slant. Note how bubbles emerge from the organisms.
2. Repeat the test on your unknown (tube H) and record your results on the Worksheet.

Oxidase Production

The production of oxidase is one of the most significant tests we have for differentiating certain groups of bacteria. For example, all the Enterobacteriaceae are oxidase-negative and most species of *Pseudomonas* are oxidase-positive. Another important group, the *Neisseria,* are oxidase producers.

Two methods are described here for performing this test. The first method utilizes the entire TSA plate; the second method is less demanding in that only a loopful of organisms from the plate is used. The two methods are equally reliable.

Materials:

TSA plate streaked with unknown and *P. aeruginosa*

oxidase test reagents (1% solution of dimethyl-*p*-phenylenediamine hydrochloride)

Whatman No. 2 filter paper

Petri dish

Entire Plate Method

Onto the TSA plate that you streaked your unknown and *P. aeruginosa,* pour some of the oxidase test reagent, covering the colonies of both organisms.

Observe that the *Pseudomonas* colonies first become **pink,** then change to **maroon, dark red,** and finally **black.** If your unknown follows the same color sequence, it, too, is oxidase-positive. Record your results on the Descriptive Chart.

Filter Paper Method

On a piece of Whatman No. 2 filter paper in a Petri dish, place several drops of oxidase test reagent. Remove a loopful of the organisms from one of the colonies and smear the organisms over a small area of the paper. The positive color reaction described above will show up within 10-15 seconds. Record your results on the Worksheet.

NITRATE REDUCTION

Many facultative bacteria are able to use the oxygen in nitrate as a hydrogen acceptor in anaerobic respiration, thus converting nitrate to nitrite. This enzymatic reaction is controlled by an inducible enzyme called *nitratase.*

$$NO_3^- + 2e^- + 2H^+ \xrightarrow{\text{nitratase}} NO_2^- + H_2O$$

Since the presence of free oxygen prevents nitrate reduction, actively multiplying organisms will use up the oxygen first and then utilize the nitrate. In culturing some organisms, it is desirable to use anaerobic methods to ensure nitrate reduction.

The nitrate broth used in this test consists of beef extract, peptone, and potassium nitrate. To test for nitrite after incubation, we use two reagents designed as A and B.

Reagent A contains sulfanilic acid and reagent B contains dimethyl-alpha-naphthylamine. In this presence of nitrite these regeants cause the culture to turn red. Negative results must be confirmed as negative with zinc dust.

Materials:

nitrate broth cultures of unknown (tube G) and test control *E. coli* (tube G^1)

nitrite test reagents (solutions A and B)

zinc dust

1. Add two or three drops of nitrite test solution A (sulfanilic acid) and an equal amount of solution B (dimethyl-alpha-naphthylamine) to the nitrate broth culture of *E. coli* (tube G^1).

   A red color should appear almost immediately, indicating that nitrate reduction has occurred.

   Caution: Since the agent in solution B is carcinogenic, avoid skin contact.

2. Repeat this procedure with your unknown (tube F). If the red color does not develop, your unknown is negative for nitrate reduction. All negative results should be confirmed as being negative as follows:

   Negative Confirmation: Add a pinch of zinc dust to the tube and shake it vigorously. If the tube becomes red, the test is confirmed as being negative. Zinc causes this reaction by reducing nitrate to nitrite; the newly formed nitrite reacts with the reagents to produce the red color.

3. Record your results in the Worksheet.

NOTES

Exercise 30 PHYSIOLOGICAL CHARACTERISTICS: HYDROLYSIS

INTRODUCTION

Many bacteria produce exoenzymes called *hydrolases,* which split complex organic compounds into smaller units. All hydrolytic enzymes accomplish this molecular splitting in the presence of water. We have already observed one example of protein hydrolysis in Exercise 47: gelatin hydrolysis by gelatinase. In this exercise we shall observe the hydrolysis of starch, casein, fat, tryptophan, and urea. Each test plays an important role in the identification of certain types of bacteria. This exercise will be performed in the same manner as the previous one, with test controls being made for comparisons.

Figure 30–1 illustrates the general procedure to be used. Three agar plates and four test tubes will be inoculated. After incubation, some of the plates and tubes will have test reagents added to them; others will reveal the presence of hydrolysis by changes that have occurred during incubation. Proceed as follows:

I. FIRST PERIOD (INOCULATIONS)

If each student is working with only one unknown, students can work in pairs to share Petri plates. Note in figure 30–1 how each plate can serve for two unknowns with the test control organism streaked down the middle. If each student is working with two unknowns, the plate will not be shared. Whether the two tubes for test controls will be shared depends on the availability of materials.

Materials:

per pair of students with one unknown each, or for one student with two unknowns:

1 starch agar plate

1 skim milk agar plate

1 spirit blue agar plate

3 urea broths

3 tryptone broths

nutrient broth cultures of *B. subtilis, E. coli, S. aureus,* and *P. vulgaris*

1. Label and streak the three different agar plates in the manner shown in figure 30–1. Note that straight-line streaks are made on each plate. Indicate, also, the type of medium in each plate.
2. Label a tube of urea broth *P. VULGARIS* and a tube of tryptone broth *E. COLI.* These will be your test controls for urea and tryptophan hydrolysis. Inoculate each tube accordingly.
3. For each unknown, label one tube of urea broth and one tube of tryptone broth with the code number of your unknown. Inoculate each tube with the appropriate unknown.
4. Incubate the plates and two test control tubes at 37° C. Incubate the unknown tubes of urea broth and tryptone broth at the optimum temperatures for the unknowns.

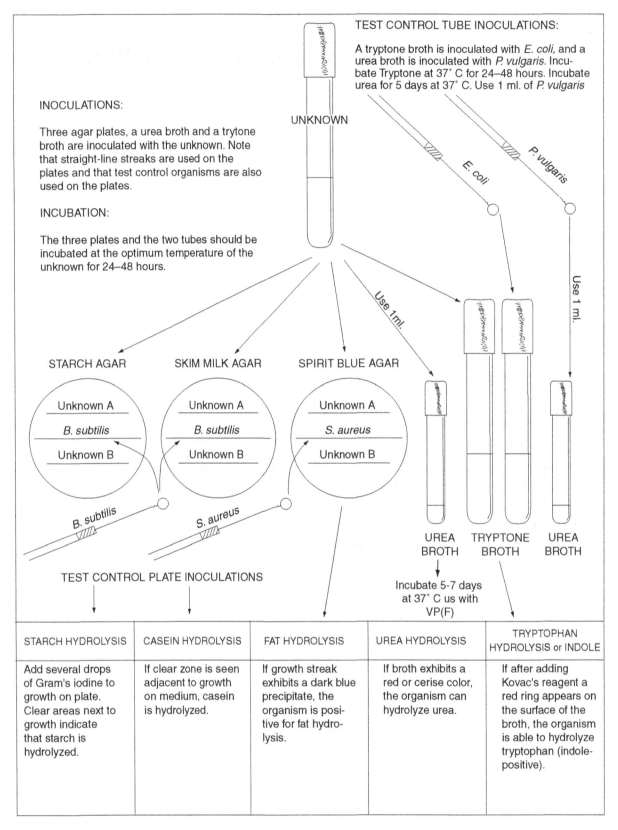

FIGURE 30–1

Procedure for doing hydrolysis tests on unknowns

II. SECOND PERIOD (EVALUATION OF TESTS)

After 24 to 48 hours incubation of unknowns and test controls, compare your unknowns with the test controls, recording all data on the Worksheet.

Starch Hydrolysis

Since many bacteria are capable of hydrolyzing starch, this test has fairly wide application. The starch molecule is a large one consisting of two constituents: amylose, a straight chain polymer of 200 to 300 glucose units, and amylopectin, a larger branched polymer with phosphate groups. Bacteria that hydrolyze starch produce *amylases* that yield molecules of maltose, glucose, and dextrins.

Materials:

Gram's iodine

starch agar culture plate

Iodine solution (Gram's) is an indicator of starch. When iodine comes in contact with a medium containing starch, it turns blue. If starch is hydrolyzed and starch is no longer present, the medium will have a **clear zone** next to the growth.

By pouring Gram's iodine over the growth on the medium, one can see clearly where starch has been hydrolyzed. If the area immediately adjacent to the growth is clear, amylase is produced.

Pour enough iodine over each streak to completely wet the entire surface of the plate. Rotate and tilt the plate gently to spread the iodine. Compare your unknown with the positive results seen along the growth of *B. subtilis*.

Casein Hydrolysis

Casein is the predominate protein in milk. Its presence causes milk to have its characteristic white appearance. Many bacteria produce the exoenzyme *caseinase,* which hydrolyzes casein to produce more soluble, transparent derivatives. Protein hydrolysis also is referred to as *proteolysis,* or *peptonization.*

Examine the streaks on the skim milk agar plates. Note that a **clear zone** exists adjacent to the growth of *B. subtilis*. This is evidence of casein hydrolysis. Compare your unknown with this positive result and record the results on the Worksheet.

Fat Hydrolysis

The ability of organisms to hydrolyze fat is accomplished with the enzyme *lipase*. In this reaction the fat molecule is split to form one molecule of glycerol and three fatty acid molecules.

The glycerol and fatty acids produced in this reaction can be used by the organism to synthesize bacterial fats and other cell components. In many instances they are even oxidized to yield energy under aerobic conditions. This ability of bacteria to decompose fats plays a role in the rancidity of certain foods, such as margarine.

Spirit blue agar contains a vegetable oil that, when hydrolyzed by most organisms, results in the lowering of the pH sufficiently to produce a **dark blue precipitate.** Unfortunately, the hydrolytic action of some organisms on this medium does not produce a blue precipitate because the pH is not lowered sufficiently.

Examine the *S. aureus* growth carefully. You should be able to see this dark blue reaction.

Compare the positive reaction of *S. aureus* with the reaction on your unknown. *If your unknown appears to be negative, hold the plate up toward the light and look for a region near the growth where oil droplets are depleted.* If you see depletion of oil drops, consider your organism to be positive for this test. Record the results on the Worksheet.

Tryptophan Hydrolysis

Certain bacteria such as *E. coli* have the ability to split the amino acid tryptophan into indole and pyruvic acid. The enzyme that causes this hydrolysis is *tryptophanase*. Indole can be easily detected with Kovacs' reagent. This test is particularly useful in differentiating *E. coli* from some closely related enteric bacteria.

Tryptone broth (1%) is used for this test because it contains a great deal of tryptophan. Tryptone is a peptone derived from casein by pancreatic digestion.

Materials:

Kovacs' reagent

tryptone broth cultures of unknown and *E. coli*

To test for indole add 10 to 12 drops of Kovacs' reagent to the *E. coli* culture in tryptone broth. **A red layer** should form at the top of the culture. Repeat the test on your unknown and record the results on the Worksheet.

Urea Hydrolysis

The differentiation of gram-negative enteric bacteria is greatly helped if one can demonstrate that the unknown can produce *urease*. This enzyme splits off ammonia from the urea molecule, as shown below. Note in the separation outline in figure 51.3 that three genera (*Proteus*, *Providencia*, and *Morganella*) are positive for the production of this hydrolytic enzyme.

Urea broth is a buffered solution of yeast extract and urea. It also contains phenol red as a pH indicator. Since urea is unstable and breaks down in the autoclave at 15 psi steam pressure, it is often sterilized by filtration. It is tubed in small amounts to hasten the visibility of the reaction.

When urease is produced by an organism in this medium, the ammonia that is released raises the pH. As the pH becomes higher, the phenol red changes from a yellow color (pH 6.8) to a **red** or **cerise color** (pH 8.1 or more).

Examine your tube of urea broth that was inoculated with *Proteus vulgaris*. Compare your unknown with this standard. *If your unknown is negative incubate the tube for a total of 7 days to check for a slow urease producer.* Record your result on the Worksheet.

NOTES

Exercise 31   PHYSIOLOGICAL CHARACTERISTICS: MISCELLANEOUS TESTS

INTRODUCTION:

There are several additional physiological tests used in unknown identification that are best grouped separately as "miscellaneous tests." They include tests for hydrogen sulfide production, citrate utilization, phenylalanine deaminization, and litmus milk reactions. During the first period, inoculations of four kinds of media will be made for these tests. An explanation of the value of the IMViC test will also be included.

I. FIRST PERIOD (INOCULATIONS)

Since test controls are included in this exercise, two sets of inoculations will be made. For economy of materials, one set of test controls will be made by students working in pairs.

Materials:
for test controls, per pair of students:
1 Kligler's iron agar deep or SIM medium
1 Simmons citrate agar slant
1 phenylalanine agar slant
nutrient broth cultures of *Proteus vulgaris*, and *Enterobacter aerogenes*

per unknown, per student:
1 Kligler's iron agar deep or SIM medium
1 Simmons citrate agar slant
1 phenylalanine agar slant
1 litmus milk

1. Label one tube of Kligler's iron agar (or SIM medium) *P. VULGARIS* and additional tubes with your unknown numbers. Inoculate each tube by stabbing with a straight wire.
2. Label one tube of Simmons citrate agar *E. AEROGENES* and additional tubes with your unknown numbers. Use a straight wire to streak stab each slant; i.e., streak the slant first, and then stab into the middle of the slant.
3. Label one tube of phenylalanine agar slant *P. VULGARIS* and the other with your unknown code number. Streak each slant with the appropriate organism.
4. With a loop, inoculate one tube of litmus milk with your unknown. (**Note:** A test control for this medium will not be made. Figure 50.2 will take its place.)
5. Incubate the unknowns at their optimum temperatures. Incubate the test controls at 37° C for 24-48 hours.

II. SECOND PERIOD (EVALUATION OF TESTS)

After 24 to 48 hours incubation, examine the tubes to evaluate according to the following discussion. Record all results on the Worksheet.

Hydrogen Sulfide Production

Certain bacteria, such as *Proteus vulgaris*, produce hydrogen sulfide from the amino acid cysteine. These organisms produce the enzyme *cysteine desulfurase*, which works in conjunction with the coenzyme pyridoxyl phosphate. The production of H_2S is the initial step in the deamination of cysteine as indicated below:

| cysteine | α amino acrylic acid | imino acid | pyruvic acid |

Kligler's iron agar or SIM medium is used here to detect hydrogen sulfide production. Both of these media contain iron salts that react with H_2S to form a **dark precipitate** of iron sulfide.

Kligler's iron agar also contains glucose, lactose, and phenol red. When this medium is used in slants it is an excellent medium for detecting glucose and lactose fermentation. SIM medium, on the other hand, can also be used for determining motility and testing for indole production.

Examine the tube of one of these media that was inoculated with *P. vulgaris*. Positive tubes have a black precipitate. Compare your unknown with this control tube and record your results on the Worksheet.

Citrate Utilization

The ability of some organisms, such as *E. aerogenes* and *Salmonella typhimurium*, to utilize citrate as a sole source of carbon can be a very useful differentiation characteristic in working with intestinal bacteria. Koser's citrate medium and Simmons citrate agar are two media that are used to detect this ability in bacteria. In both of these synthetic media sodium citrate is the sole carbon source; nitrogen is supplied by ammonium salts instead of amino acids.

Examine the test control slant of this medium that was inoculated with *E. aerogenes*. Note the distinct **Prussian blue color change** that has occurred. Record your results on the Worksheet.

Phenylalanine Deamination

A few bacteria, such as *Proteus, Morganella,* and *Providencia,* produce the deaminase *phenylalanine,* that deaminizes the amino acid phenylalanine to produce phenylpyruvic acid (PPA). This characteristic is used to help differentiate these three genera from other genera of the Enterobacteriaceae. The reaction is as follows:

Proceed as follows to test for the production of phenylpyruvic acid which is evidence that the enzyme pheny-

| PHENYLALANINE | | PHENYLPYRUVIC ACID |

lalanase has been produced:

Materials:

dropping bottle of 10% ferric chloride

Allow 5-10 drops of 10% ferric chloride to flow down over the slants of the test control (*P. vulgaris*) and your unknowns. To hasten the reaction, use a loop to emulsify the organisms into solution. **A deep green color** should appear on the test control **slant** in 1-5 minutes. Compare your unknown with the control and record your results on the Worksheet.

The IMViC Tests

In the differentiation of *E. aerogenes* and *E. coli,* as well as some other related species, four physiological tests have been grouped together into what are called the IMViC tests. The *I* stands for indole; the *M* and *V* stand for methyl red and Voges-Proskauer tests; *i* simply facilitates pronunciation; and the *C* signifies citrate utilization. In the differentiation of the two coliforms *E. coli* and *E. aerogenes,* the test results appear as charted below, revealing completely opposite reactions for the two organisms on all tests.

	I	M	V	C
E. Coli	+	+	-	-
E. aerogenes	-	-	+	+

The significance of these tests is that when testing drinking water for the presence of the sewage indicator *E. coli,* one must be able to rule out *E. aerogenes,* which has many of the morphological and physiological characteristics of *E. coli.* Since *E. aerogenes* is not always associated with sewage, its presence in water would not necessarily indicate sewage contamination.

If you are attempting to identify a gram-negative, facultative, rod-shaped bacterial organism, group these series of tests together in this manner to see how your unknown fits this combination of tests.

Litmus Milk Reactions

Litmus milk contains 10% powdered skin milk and a small amount of litmus as a pH indicator. When the medium is made up, its pH is adjusted to 6.8. It is an excellent growth medium for many organisms and can be very helpful in unknown characterization. In addition to revealing the presence or absence of fermentation, it can detect certain proteolytic characteristics in bacteria. A number of facultative bacteria with strong reducing powers are able to utilize litmus as an alternative electron acceptor to render it colorless. Since some of the reactions take 4 to 5 days to occur, the cultures should be incubated for at least this period of time; they should be examined every 24 hours, however. Look for the following reactions:

Acid Reaction Litmus becomes pink. Typical of fermentative bacteria.

Alkaline Reaction Litmus turns blue or purple. Many proteolytic bacteria cause this reaction in the first 24 hours.

Litmus Reduction Culture becomes white; actively reproducing bacteria reduce the O/R potential of medium.

Coagulation Curd formation. Solidification is due to protein coagulation. Tilting tube at 45° will indicate whether or not this has occurred.

Peptonization Medium becomes translucent. It often turns brown at this stage. Caused by proteolytic bacteria.

Ropiness Thick, slimy residue in bottom of tube. Ropiness can be demonstrated with sterile loop.

Record the litmus milk reactions of your unknown on the Worksheet.

Laboratory Report

Complete Laboratory Report 48-50, which reviews all physiological tests performed in the last three exercises.

WORKSHEET FOR UNKNOWN BACTERIA

Unknown Number: _____

Name _____

Bacterial Identification: _____

MORPHOLOGICAL AND CULTURAL CHARACTERISTICS

Cell morphology: _____

Arrangement: _____

Colonial morphology:

Trypticase soy agar: _____

Agar slant: _____

Gram's Stain _____

Motility: _____

Spores: _____

Capsules: _____

Trypticase soy broth: _____

Gelatin Stab: _____

Oxygen requirements: _____

PHYSIOLOGICAL CHARACTERISTICS TESTS RESULTS

Glucose _____

Lactose _____

Sucrose _____

Mannitol _____

Gelatin Liquefaction _____

Starch _____

Casein _____

Fat _____

PHYSIOLOGICAL CHARACTERISTICS TESTS RESULTS

Indole _____

Methyl Red _____

V-P (acetylmethylcarbinol) _____

Citrate Utilization _____

Nitrate Reduction _____

Hydrogen Sulfide Production _____

Urease _____

Catalase _____

Oxidase _____

Phenylalanase _____

LITMUS MILK REACTION:

Acid _____

Alkaline _____

Reduction _____

Coagulation _____

Hydrolysis _____

OTHER OBSERVATIONS:

CONTROL OF MICROBIAL GROWTH

Exercise 32   EFFECTS OF TEMPERATURE ON GROWTH

INTRODUCTION

Temperature is a physical factor which affects the growth of bacteria. Overall, bacteria can grow over a temperature range of −5°C to 80°C, but most species can only grow within a range of 30°C. This narrower range is dictated by the sensitivity of the enzymes, membranes, ribosomes, and other cellular components to temperature. Each bacterial species has its own set of **cardinal temperature points.** These are the minimum, maximum, and optimum growth temperatures. The **minimum growth temperature** is the lowest temperature at which the organism will grow. If the temperature extends below this temperature, enzyme activity is inhibited and the bacteria become metabolically inactive, so that growth is negligible or nonexistent. The **maximum growth temperature** is the highest temperature at which growth occurs. At temperatures above this temperature, enzymes are denatured and bacteria die. Bacterial growth is the most rapid at the **optimum growth temperature.** This is the temperature when bacteria are multiplying at their quickest rate. This temperature may not coincide, however, with the most ideal temperature for other cellular mechanisms. Typically, organisms are found growing in an environment that supports the optimum growth temperature requirements for the organisms living in that environment.

Based on their preferred growth temperatures, bacteria can be classified into one of three major groups. **Psychrophiles** are bacteria that grow within a temperature range of −5°C to 20°C and have an optimum growth temperature of approximately 15°C. Some psychrophiles are so sensitive to temperature that they cannot grow when heated to 20°C. These are considered to be **obligate psychrophiles. Facultative psychrophiles** grow best below 20°C, but are capable of growing above this temperature. Some psychrophiles are the culprits for spoiling refrigerated foods. **Mesophiles** grow within a temperature range of 20°C to 45°C. Within the mesophiles, there are two distinct groups. Plant saprophytes have an optimum growth temperature between 20°C and 30°C. Other organisms in the mesophile group prefer to grow in the bodies of warm-blooded hosts and, therefore, have an optimum growth temperature between 35°C and 40°C. Because this is the same temperature as the human body, organisms within this group can infect humans. **Thermoduric** organisms are those which can survive exposure to high temperatures for short periods of time, even though they normally grow as mesophiles. Finally, **thermophiles** include bacteria which grow best at temperatures above 45°C. Organisms known as **facultative thermophiles** are able to grow at 37°C, but have optimum growth temperatures of 45°C to 60°C. Conversely, obligate thermophiles will only grow at temperatures above 50°C and have optimum growth temperatures above 60°C.

Materials:

4 Tryptic soy agar (TSA) plates

Bunsen burner

Inoculating loop

Permanent marking pen

Cultures:

Bacillus stearothemophilus

Escherichia coli

Pseudomonas fluorescens

Staphylococcus aureus

Procedure

1. This procedure will be performed by each group, so groups begin by carefully reviewing the procedure and by dividing the labor between group members.

2. Each group will receive stock cultures of *Escherichia coli*, *Pseudomonas fluorescens*, *Staphylococcus aureus*, and *Bacillus stearothermophilus*.

3. Groups should receive four tryptic soy agar (TSA) plates, one for each temperature being tested. These plates are all identical. The differences will be the temperature at which the plates are incubated.

4. Groups should divide the bottom of each of the four plates into four quadrants by using a wax pencil to draw a set of perpendicular lines on the plate. See Figure 32–1.

5. Each organism should be spot-inoculated into one quadrant on each of the four plates.

6. The plates should be carefully labeled with the customary information and should also include the incubation temperature as well as the organism in each quadrant.

7. The plates will be incubated for 48 hours, one each at the following temperatures:-

    4°C

    25°C

    37°C

    55°C

8. The following laboratory period, presence or absence of growth for each organism should be recorded for each temperature and determinations as to the classification of each organism based on the optimum growth temperature should be made.

9. Record your results on the Worksheet.

10. Dispose of the plates by placing them in the biohazard bag in your laboratory.

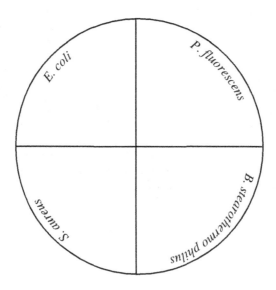

FIGURE 32–1
Example of how to divide one TSA plate for temperature testing.

Name _____   Date _____

WORKSHEET

	Escherichia coli	Pseudomonas fluorescens	Staphylococcus aureus	Bacillus stearothermophilus
4°C Growth				
25°C Growth				
37°C Growth				
55°C Growth				
Temperature Classification				

QUESTIONS

1. List five physical factors that affect bacterial growth.

_____

_____

_____

_____

_____

2. Differentiate between the following:

Thermophile:

Mesophile:

Psycrophile:

3. What is the optimal temperature for each of the bacteria you tested?

_____

_____

_____

_____

_____

Exercise 33   EFFECTS OF ULTRAVIOLET LIGHT

INTRODUCTION

Radiation, or radiant energy, is transmitted from the sun and other natural and man-made sources to the earth. Radiation differs in both wavelength and energy, with the shorter wavelengths containing more energy. Some wavelengths of light are capable of causing **mutations** in living cells. A **mutation** is simply a change in the DNA sequence. The effects of a mutation can range from being totally harmless to causing the death of the cell. **X-rays** (wavelengths from 0.1 to 40 nm) and **gamma rays** are both forms of **ionizing radiation.** These types of rays have short wavelengths and are said to be "ionizing" because of their ability to eject electrons from atoms or molecules. These electrons directly damage the DNA of the cell and cause peroxides to form, which can also be harmful to the cell. Some wavelengths of radiant energy, those in the **nonionizing wavelengths,** are actually essential for biochemical processes. However, even these wavelengths are harmful to some cells, including bacteria and viruses.

Wavelength (Nanometers)

10^{-5}	10^{-3}	10^{-1}	10	400	700	10^8	10^{10}
Cosmic Rays	Gamma Rays	X-Rays	Ultraviolet Rays	←Violet Red→	Infrared Rays	Radio Waves	

Visible

FIGURE 33–1
The various wavelengths of radiation.

One form of nonionizing radiation, known as **ultraviolet radiation,** has wavelengths ranging from 40 to 390 nm. The wavelength that range between 260 and 265 nm are the most harmful. Ultraviolet radiation exerts its effects on the DNA of all types of cells and causes covalent bonds to form between adjacent thymine bases (pyrimidine bases). The formation of these **thymine dimers** is a lethal mutation because they inhibit correct replication of the DNA during cellular reproduction. Bacteria and other organisms have developed mechanisms of repairing the damage caused by exposure to ultraviolet light. When bacteria containing thymine dimers are exposed to visible light, the enzyme **pyrimidine dimerase** is activated. This enzyme acts to split the thymine dimer in a reaction known as **light-repair** or **photoreactivation.** DNA replication can then proceed normally and the bacteria do not die. It is important to note that exposure of human skin cells to ultraviolet radiation, such as that found in sunlight, causes high numbers of thymine dimers to form. Human skin cells do have pyrimidine dimerase, but it is the unrepaired mutations that can lead to skin cancers.

Because ultraviolet radiation can be lethal to microorganisms, this type of radiation is routinely used to control

From *Exercises for the Microbiology Laboratory* by Elizabeth Fish McPherson. Copyright © 2001 by Elizabeth Fish McPherson. Reprinted by permission of Kendall/Hunt Publishing Company.

thymine dimer

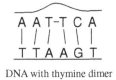

normal DNA   DNA with thymine dimer

normal DNA vs DNA containing
a thymine dimer

microbial growth in the air, hospital rooms, nurseries, operating rooms, and cafeterias. Another use for ultraviolet radiation is to sterilize materials such as vaccines, serum, and toxins, which may not withstand the conditions of the autoclave. In some areas ultraviolet radiation is used in place of chlorine to purify wastewater and drinking water. Ultraviolet light does not penetrate glass, plastic, paper, or some other materials, and only sterilizes the area directly in contact with the light. UV light does not bend around corners or go underneath benches. This type of radiation does not affect organisms protected by solids and coverings. Another potential problem with ultraviolet light is that it can damage the eyes and cause burns and skin cancer following moderate to prolonged exposure.

In today's exercise, we will examine the effects of ultraviolet light on a bacterial culture. Two different exposure times will be utilized to determine if length of exposure to ultraviolet light affects the bacteria. Additionally, two different incubation conditions will be employed to examine if the bacteria are capable of light repair.

Materials:

16 Nutrient agar plates

Ultraviolet lamp, or UV exposure box

Sterile cotton swabs

Index cards ($5'' \times 3''$)

Timer

Permanent marking pen

Safety glasses

Cultures:

Serratia marcescens

Bacillus megaterium

Procedure:

1. Refer to Table 33-1 to determine the organism and exposure times you will be working with.
2. Label the bottom of each streaked plate with one of each exposure time. Draw a line dividing the bottom of each plate into two sections. One half of the plate will be irradiated, and the other half will be covered.

3. Dip a sterile cotton swab into one of the bacterial cultures. Streak the surface of the agar with the swab, and be sure to cover the entire plate as evenly as possible. Rotate the plate 90 degrees and streak the surface again. It is important to spread the bacteria evenly on the plate. When you are done with the swab, put the swab in the beaker with germicide. **See Fig. 33–2.**

Step 1   Step 2

FIGURE 33–2

From *Fundamental Microbiology for the Health Care Sciences,* Fourth Edition by Frank A. Hartley, Walter Hoeksema, and Michael Ryan. Copyright © 2001 by Kendall/Hunt Publishing Company. Reprinted by permission.

4. Repeat step 3 with each plate, using a fresh cotton swab with each bacteria.
5. Remove the cover of the first plate to be used. Place a card over the part of the plate that will not be irradiated. If necessary, secure the card to the side of the Petri plate with a small piece of tape. Expose the plate for the designed time. *Be sure to leave the cover off the plate.*
6. After exposure, replace the cover.
7. Repeat exposures with each Petri plate.
8. Expose plate # 8 with its lid on. DO NOT REMOVE THE LID.
9. Store the plates in an inverted position.
10. Incubate plates at 37°C. After incubation, examine all plates for the presense of bacteria. Compare the bacteria on the exposed and covered portions of the plate.

TABLE 33–1

		Exposure Times						Lid On
S. marcescens	Plate 1	2	3	4	5	6	7	8
	10 sec	20 sec	40 sec	80 sec	2.5 min	5 min	10 min	10 min
B. megaterium	Plate 1	2	3	4	5	6	7	8
	1 min	2 min	4 min	8 min	15 min	30 min	60 min	60 min

Name _____   **Date** _____

WORKSHEET

You will be tabulating your results by estimating the number of surviving colonies, after exposure to UV light. Enter your results in the table below as follows.

Substantial survival	$+++$
Moderate survival	$++$
Low survival (1-3 colonies)	$+$
No survival	0

	Exposure Time							**With Lid On**
S. marcescens	10 sec	20 sec	40 sec	80 sec	2.5 min	5 min	10 min	10 min
Survival								
B.megaterium	1 min	2 min	4 min	8 min	15 min	30 min	60 min	60 min
Survival								

QUESTIONS

1. What length of time is required for the destruction of spore-forming bacteria such as *Bacillus megaterium*?
2. How does UV radiation kill bacteria?
3. Sunlight includes the most germicidal wavelength of UV radiation. Why doesn't this wavelength reach the earth's surface?
4. Of what value might pigment production be in protecting a bacterium against UV radiation?
5. Describe the relationship between UV radiation exposure time and the number of bacteria killed?

Exercise 34   ANTIMICROBICS

INTRODUCTION

Chemotherapeutic agents are chemical agents that inhibit or prevent microbial growth, are used internally, and may be natural or synthetic in nature. Historically speaking, in 1928 Alexander Fleming discovered that growth of *Staphylococcus aureus* was inhibited in the area surrounding a colony of the mold *Penicillium notatum*. The active inhibitory compound was later called penicillin, and the mechanism of inhibition was called antibiosis. The term antibiotic is derived from this word. An **antibiotic** is defined as a substance that is produced by a microorganism that in small amount is inhibitory for other microorganisms. Synthetic drugs, such as the sulfa drugs, are therefore technically not antibiotics, but this distinction is generally overlooked.

Many different types of antibiotics have been discovered, most of which are produced by species of the filamentous bacteria *Streptomyces*. Bacteria in the genus *Bacillus* produce some antibiotics, while others are produced by the fungal genera *Cephalosporium* and *Penicillium*. **Broadspectrum antibiotics** are effective against a large range of both Gram positive and Gram negative bacteria. **Narrow spectrum antibiotics** are effective against only a restricted number of bacteria. Common thought dictates that because the causative agent in an infection is not normally known at the time an antibiotic is prescribed, a broad-spectrum antibiotic would be the best choice. This is not necessarily always the case because the broad-spectrum drugs affect the normal flora of the host. It is generally best to receive a drug with a more narrow spectrum of activity, so that such side effects can be prevented. However, this is not always practical because it takes several days to identify a microorganism.

When a physician prescribes an antibiotic, the intention is for the drug to kill the organism(s) causing the infection and not cause harm to the individual. Antibiotics have five different mechanisms by which they kill bacteria. An antibiotic may interfere with the synthesis of the peptidoglycan layer in bacteria or may destroy this layer outright. Such antibiotics have no effect on eukaryotic cells because eukaryotes do not have peptidoglycan in their cell walls. The cytoplasmic membranes of bacteria are susceptible to the action of a class of antibiotics, but because prokaryotic and eukaryotic cytoplasmic membranes are similar, these drugs also can harm mammalian cells. Recall that ribosomes in prokaryotes are 70S, while those in eukaryotes are 80S. This difference is exploited by antibiotics that interfere with protein synthesis. Although the physical structure and composition of nucleic acids is identical between prokaryotic and eukaryotic cells, the enzymes that aid in DNA replication and transcription are different. Thus, the class of antibiotics that inhibits nucleic acid synthesis is selectively toxic for prokaryotes. Finally, an antibiotic may function as an antimetabolite in the synthesis of the vitamin folic acid which is essential to the growth of bacteria. These drugs are incorporated in various steps in the pathway that synthesizes folic acid. When these drugs are incorporated, synthesis of folic acid ceases and microbial growth soon stops. Human do not synthesize folic acid, but rather obtain it in our diet, so again, this class of antibiotics is selectively toxic.

The most widely used test to determine the susceptibility of an organism to an antibiotic is the **Kirby-Bauer test.** In this method, antibiotics are impregnated onto paper disks and are them placed on a TSA plate swabbed with the organism in question. During the subsequent incubation, the antibiotics diffuse from the disks into the agar from an area of high concentration to an area of low concentration. In some situations, a **zone of inhibition** will appear around the disk. A measurement of the diameter of the zone of inhibition is made and this measurement compared to a table that indicates the measurements for susceptible, intermediate, and resistant organisms. Just because an antibiotic produces a zone does not automatically indicate that the organism is sensitive to the antibiotic. Comparison of the measured value with a data table must be performed. The results are often insufficient for clinical purposes, but this is simple, inexpensive, and does have considerable value in medical practices.

Materials:
Three TSA plates
Disk dispenser (BBL or Difco)
Cartridges of disks (BBL or Difco)
Bunsen burner
Forceps
Sterile cotton swabs
Permanent marking pen
Metric Ruler

Cultures:
Staphylococcus aureus
Pseudomonas aeruginosa
Escherichia coli

Procedure:
1. This procedure will be performed by each group, so groups should begin by carefully reviewing the procedure and by dividing the labor between group members.
2. Each group will receive stock cultures of *Staphylococcus aureus, Pseudomonas aeruginosa,* and *E. coli.*
3. Each group will receive three TSA plates.
4. Using a sterile cotton swab dipped in one stock bacterial culture, swab the entire surface of the plate with the organism. This is best done by swabbing in at least two directions with the swab.
5. Repeat the above procedure with the remaining organisms.
6. Discard the swabs.
7. Dispense the antibiotics onto the plate with the multiple disk dispenser. Be sure to remove the lid of the plate period to dispensing the antibiotics.
8. Make sure contact is made between the antibiotic disks and the agar by gently pressing the disk onto the surface of the plate with alcohol-flamed forceps. Do not press the disk onto the agar and do not move the disk once it is placed on the agar.
9. Label the plates carefully.
10. The plates will be incubated for 24 hours at 37°C. The plates will not be inverted during this incubation.
11. The following laboratory period, measure the diameter of the zone of inhibition (see Fig. 34–1) around each antibiotic disk and compare this value to Table 34–1 to determine if the organism is susceptible, intermediate, or resistant to the particular antibiotic. Record this information on the Worksheet.

TABLE 34–1
Antibiotic Zones of Inhibition

Antibiotic	Inhibition Zone (mm)			
	Disk Concentration	Resistant	Intermediate	Susceptible
Ampicillin	AM 10 mcg	28 or less		29 or more
Chloramphenicol	C 30 mcg	12 or less	13-17	18 or more
Erythromycin	E 15 mcg	13 or less	14-17	18 or more
Kanamycin	K 30 mcg	13 or less	14-17	18 or more
Neomycin	N 30 mcg	12 or less	13-16	17 or more
Novobiocin	NB 30 mcg	17 or less	18-21	21 or more
Penicillin for Staph.	P 10 mcg	20 or less	21-28	29 or more
Streptomycin	S 10 mcg	11 or less	12-14	15 or more
Tetracycline	TE 30 mcg	14 or less	15-18	18 or more

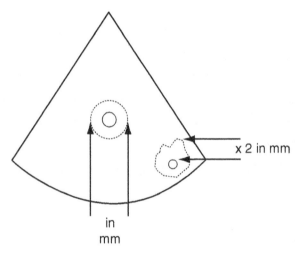

FIGURE 34–1
Measurement of the Zone of Inhibition.

Name _____   **Date** _____

WORKSHEET

Antibiotic Name	Staphylococcus aureus		Pseudomonas aeruginosa		E. coli	
	Zone Diameter	R, I, or S?	Zone Diameter	R, I, or S?	Zone Diameter	R, I, or S?
Ampicillin						
Chloramphenicol						
Erythromycin						
Kanamycin						
Neomycin						
Novobiocin						
Penicillin						
Streptomycin						
Tetracycline						

QUESTIONS

1. What is meant by a broad-spectrum antibiotic? Which antibiotics used in this lab appear to be broad-spectrum?

   _____

   _____

   _____

2. Is penicillin a broad-spectrum antibiotic? Why or why not?

   _____

   _____

   _____

   _____

3. Is an antiseptic the same thing as an antibiotic? Why?

   _____

   _____

   _____

   _____

4. Which antibiotic is the most effective and the least effective against the G− an the G+ bacteria used in this lab?

   _____

   _____

   _____

   _____

5. Consider the following situation.

   A pharmaceutical company has developed a chemical agent that inhibits the synthesis of the bacterial lipopolysaccharide, Lipid A.

   (a) Do you predict that this chemical would be more effective against gram-positive or gram-negative bacteria? Explain.

   _____

   _____

   _____

   _____

   (b) Besides toxicity to bacteria, what other properties of this chemical will determine whether or not it can be used as an antibiotic?

   _____

   _____

   _____

Adapted from *Microbiology Laboratory Manual* by Barnes and Brand and *Basic Concepts in Microbiology Lab* by J. D. Hendrix.

NOTES

Exercise 35 EFFECTS OF ANTISEPTICS AND DISINFECTANTS

INTRODUCTION:

We have examined many different ways to encourage the growth of microbes. Now, we are going to investigate methods of inhibiting or preventing microbial growth. There are both physical and chemical processes for preventing microbial growth. **Physical methods** of microbial control include regulation of some of the physical factors required for bacterial growth that we examined in Exercises 16-20. Changes in temperature, moisture, pH, osmotic pressure, and oxygen affect the growth of microorganisms. As we learned, each microorganism has its own specific growth requirements. If conditions are manipulated such that these requirements cannot be met, the organisms will not grow. We will also examine another physical factor which affects microbial growth in Exercise 44. **Chemical factors** used to control microbial growth include disinfectants, antiseptics, and chemotherapeutic agents. **Disinfectants** are chemical agents used on inanimate objects to lower the level of microbes on a surface. In contrast, **antiseptics** are chemicals used on living tissues to decrease the number of microbes on the surface. Finally, **chemotherapeutic agents** are chemical agents that are used internally and may be natural or synthetic in nature. All of these physical and chemical methods of bacterial growth inhibition are evaluated on their abilities to deter microbial growth. **Microbicidal agents** are those agents that result in microbial death, while **microbiostatic agents** are those agents that cause a temporary inhibition in microbial growth. Microbial growth resumes when the microbiostatic agent is removed.

The modes of action of these physical and chemical agents vary. Some of the effects of these agents include denaturation and inactivation of enzymes, interference with the structure and/or function of the bacterial DNA, inhibition of protein function, disruption of the cytoplasmic membrane, disruption of the cell wall, and dissolution of cellular lipids. As we have learned in previous exercises, enzymes, DNA, proteins, the cytoplasmic membrane, the cell wall, and lipids are all essential for life. Any agent that interferes with those components disrupts the normal functioning of the cell in such a way that inhibition of growth or cell death results. In order to correctly select and apply the physical or chemical agent, it is necessary to know the mode of action of the agent.

In this exercise we will focus on the antiseptics and disinfectants. Some antiseptics and disinfectants include phenol and phenolics, halogens, alcohols, heavy metals and their compounds, surfactants, and quaternary ammonium compounds. **Phenol** is often used in throat lozenges for its local anesthetic effect. The concentration of phenol is not high enough in these lozenges to cause any antimicrobial effect, but in throat sprays where the concentration is above 1%, the phenol does have an antibacterial effect. **Phenolics** are composed of a chemically altered molecule of phenol in conjunction with a soap or detergent. The active ingredient in pHisoHex (a facial cleanser) is a phenolic compound. **Halogens** include iodine and chloride, either alone or combined with organic or inorganic compounds. Iodine is one of the oldest and most effective of all antiseptics. One common commercial preparation that includes iodine is Betadine®, which is used to cleanse wounds in both humans and animals. A liquid form of chlorine gas is used to disinfect drinking water, water in swimming pools, and raw sewage. **Alcohols** are probably the most common method of ridding living and nonliving surfaces of bacteria and fungi. As we have all experienced, the skin is swabbed with 70% ethanol prior to receiving an injection. Alcohols are useful for this purpose because they exert their effects immediately, then evaporate and leave no residue behind. **Heavy metals,** such as 1% silver nitrate solution, are used as antiseptics. In past years many states required silver nitrate drops to be placed in the eyes of newborns to guard against *Neisseria gonorrhoeae* infection. However, antibiotics have now replaced the silver nitrate solution. Another heavy metal, copper, has been used to inhibit the growth of green algae in swimming pools and fish tanks. **Surfactants** are surface-active agents such as soaps and detergents. These compounds help to break through the oily layer on the skin, a process known as emulsification. Once this layer has been disrupted, the manual action of scrubbing helps remove debris and microbes. Surfactants alone are not effective antiseptics but it is through the scrubbing procedure that these agents work effectively. **Quaternary ammonium compounds (quats)** are the most widely used of all disinfectants. The disinfectant solution we use in the laboratory (quat 50) is one of these compounds. These agents are highly effective against Gram positive bacteria, fungi, and viruses. However, they are markedly less effective against Gram negative bacteria.

Several items must be considered when choosing an antiseptic or disinfectant for use. The concentration of the compound will affect its action. Contrary to common thought, higher concentrations are not necessarily better. The nature of the material being treated affects the action of the compound. The presence of organic material, which is present in vomit or feces, and the pH can influence the effectiveness of the compound. The ability of the compound to easily make contact with the microbes must be considered. A compound will act more efficiently if there is direct contact between it and the microbes. Finally, the temperature of the environment or the compound affects killing. Generally, the higher the temperature, the more effective the action of the compound.

In today's exercise we will compare the effectiveness of various antiseptics and disinfectants against Gram positive and Gram negative bacteria. Plates will be swabbed with bacteria and filter paper discs dipped in various compounds will be placed on the plates. If the organism on the plate is susceptible to the compound, a zone of inhibition will appear around the disc. The diameter of this zone will be measured and then compared with those obtained from the other organisms to rate the effectiveness of the compound. In this exercise, we will denote a chemical as effective if it produces a zone of inhibition.

From *Exercises for the Microbiology Laboratory* by Elizabeth Fish McPherson. Copyright © 2001 by Elizabeth Fish McPherson. Reprinted by permission of Kendall/Hunt Publishing Company.

Materials:
Eight TSA plates
Sterile cotton swabs
Forceps
95% ethanol
Sterile paper disks
Metric rulers
Phenol
Bleach (10%)
Listerine
Iodine
Disinfectant (Lysol)
Ethanol (70%)
Silver nitrate solution
Copper nitrate solution

Cultures:
Escherichia coli
Pseudomonas aeruginosa
Staphylococcus aureus
Bacillus megaterium

Procedure:
1. Each group should obtain 8 TSA plates.
2. Use a marker to divide the outside bottom of a TSA plate into quarters.
3. Label one quarter for each of the disinfectants. Also label the bottom of the dish with your initials, lab section, and the name of the microbe assigned to you.

From *Fundamental Microbiology for the Health Care Sciences*, Fourth Edition by Frank A. Hartley, Walter Hoeksema, and Michael Ryan. Copyright © 2001 by Kendall/Hunt Publishing Company. Reprinted by permission.

4. Dip a sterile cotton swab into the broth culture of your microbe. Roll the swab around the inside top portion of the culture tube to squeeze out excess fluid.

5. Use the swab to lawn-streak the surface of the medium in the labeled TSA dish (see Figure 35–1). Discard this swab.

6. Use a flamed and cooled forceps to pick up a sterile disk. Dip the disk edge into one of the disinfectants and let the disinfectant soak up into the disk.

7. Tap the disk on the edge of the vial to eliminate excess fluid, and then place the disk onto the inoculated TSA at the center of the appropriately labeled quarter.

8. Repeat step 5 and 6 for the remaining disinfectants. Flame and cool the forceps between each disk.

9. Press each disk gently but firmly with the forceps so each disk makes firm contact with the medium. Leave the dish right side up and incubate at 37°C.

10. During the next class period, examine each of the TSA plates and measure the diameter of the zone of inhibition around each disk. Enter your data into Table 35–1.

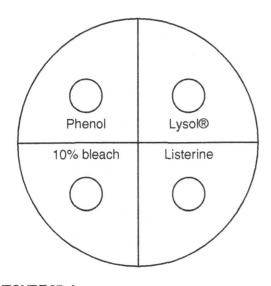

FIGURE 35–1

TABLE 35–1

	Iodine	70% Ethanol	Bleach	Phenol	Listerine	Lysol	Silver Nitrate	Copper Nitrate
Pseudomonas aeruginosa								
Escherichia coli								
Staphylococcus aureus								
Bacillus megaterium								

QUESTIONS

1. List six factors that might influence the antiseptic or disinfectant qualities of a particular chemical agent.

   a.

   b.

   c.

   d.

   e.

   f.

2. Which compound is more effective against Gram positive organisms?

   _____

   _____

   _____

   _____

3. Which compound is more effective against Gram negative organisms?

   _____

   _____

   _____

   _____

4. Which compound is the most effective overall?

   _____

   _____

   _____

   _____

NOTES

Exercise 36 KINGDOM FUNGI

I. INTRODUCTION

The **fungi** are a large group of eukaryotic organisms that lack chlorophyll and utilize absorption as their means of nutrition. The unicellular fungi are called **yeasts. Molds** are multicellular filamentous organisms such as mildews, rusts, and smuts. **Fleshy fungi** are multicellular, and include mushrooms, puffballs, and coral fungi. All fungi are **heterotrophs** that require organic compounds made by other organisms for energy and carbon. The majority of fungi feed on dead organic matter. These are called **saprophytes.** By using **extracellular enzymes,** fungi are the primary decomposers of the hard parts of plants. Other fungi are **parasitic** and obtain their nutrients from a living host.

The fungal body (**thallus**) consists of long filaments of cells called **hyphae** (singular, **hypha**). Some phyla possess hyphae that contain crosswalls called **septa,** which divide the hyphae into distinct cells with one nucleus. Other hyphae contain no septa and appear as long, continuous cells with many nuclei. These hyphae are called **coenocytic.** Hyphae grow and intertwine to form a mass called a **mycelium.** The **vegetative mycelium** is the portion of mycelium involved in nourishment. The portion concerned with reproduction is called the **aerial mycelium.**

Reproduction in fungi can be sexual or asexual. Yeasts reproduce by a process called **budding.** The parent cell forms a bud on its outer surface. As the bud elongates, the parent's cell's nucleus divides, and one nucleus migrates into the bud. The bud will eventually break away.

Another reproductive strategy is **spore** (sporangia) formation. Spores are formed form the aerial mycelium in a number of ways. **Asexual spores** are formed from the aerial mycelium of a single individual. After germination of the spore, the new organism is genetically identical to the parent. The different types of asexual spores are shown in Figure 36–1. **Sexual spores** are formed from the union of nuclei from two genetically different individuals. The different types of sexual spores are shown in Figure 36–2. Fungi are classified into four phyla, depending upon the type of the sexual spore that is present. The phyla are the **Zygomycota, Ascomycota, Basidiomycota,** and the **Deuteromycota** (or *Fungi Imperfecti*). The Deuteromycota are "imperfect" because no sexual spores have yet been identified or found for individuals in this grouping.

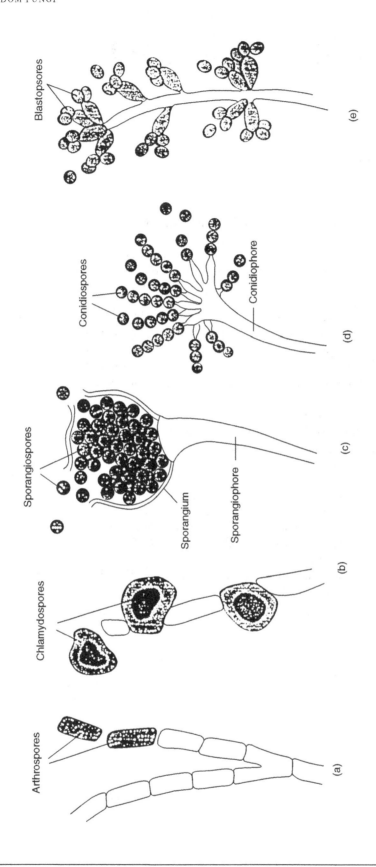

FIGURE 36–1

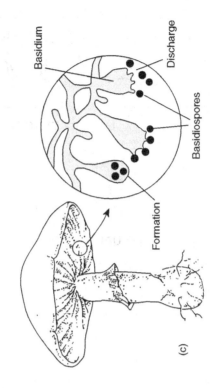

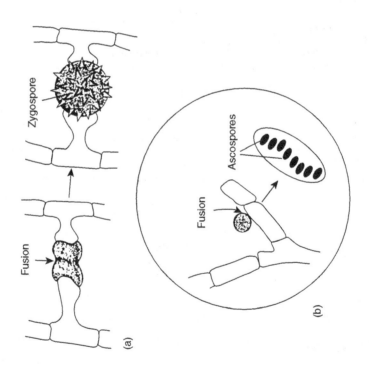

FIGURE 36–2

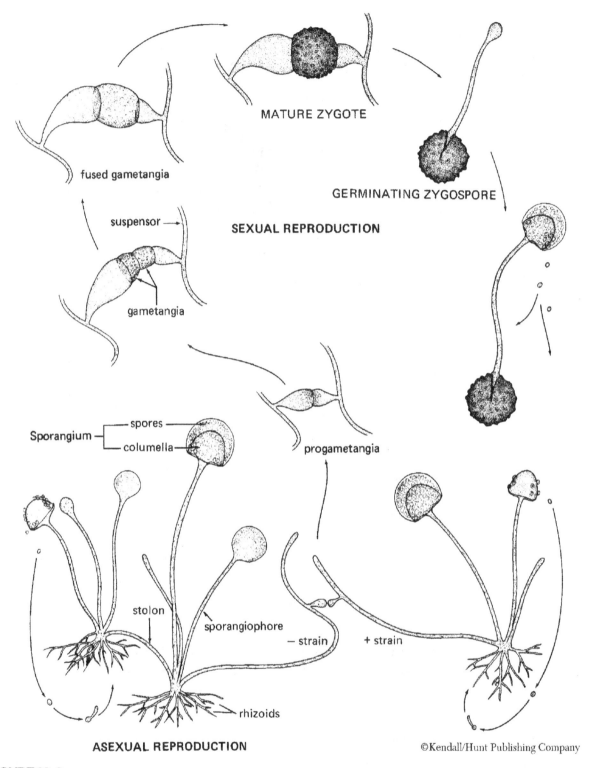

MATURE ZYGOTE

GERMINATING ZYGOSPORE

fused gametangia

suspensor →

SEXUAL REPRODUCTION

gametangia

Sporangium — spores
— columella

progametangia

stolon

sporangiophore

— strain   + strain

rhizoids

ASEXUAL REPRODUCTION

©Kendall/Hunt Publishing Company

FIGURE 36–3

II. PHYLUM ZYGOMYCOTA

The **zygomycota** includes the common bread molds, water molds, potato blight, and others. They are saprophytic and have coenocytic hyphae. The example you will be studying is *Rhizopus nigricans,* the black bread mold. The asexual spores of *Rhizopus* are **sporangiospores** and are located on asexual reproductive structures called sporangiophores (Figure 36–3). The spores inside the sporangium have a dark color, hence the descriptive name given to *Rhizopus.* The sexual spores are **zygospores.**

The mycelium is differentiated into several specialized structures. The **rhizoids** are branches that extend into the substrate and help in anchoring the fungus, and in absorption. **Stolons** connect different sections of hyphae. You should be able to observe all these structures in lab.

Procedure:

1. Obtain *Rhizopus* that has been grown in a Petri dish. The dish is sealed. DO NOT open it.

2. Using a dissecting microscope observe the hyphal filaments making up the mycelium.

3. List the structures that you can identify:

_____

_____

4. Obtain a prepared slide of *Rhizopus.* Find a section from which you can draw the major structures of the fungus. Label each structure.

Asexual Reproduction   Sexual Reproduction

III. PHYLUM ASCOMYCOTA

The **Ascomycota,** or sac fungi, include the yeasts, morels, truffles, and some molds. Yeasts are used commercially for producing baked products, and beer and wine. Morels and truffles are edible fungi. However, there are also harmful parasitic forms such as Dutch Elm Disease and Chestnut Blight. Ascomycetes are called "sac fungi" because their sexual spores are **ascospores,** produced in a sac, or **ascus.** Their asexual spores are usually **conidiospores** (Figure 36–4).

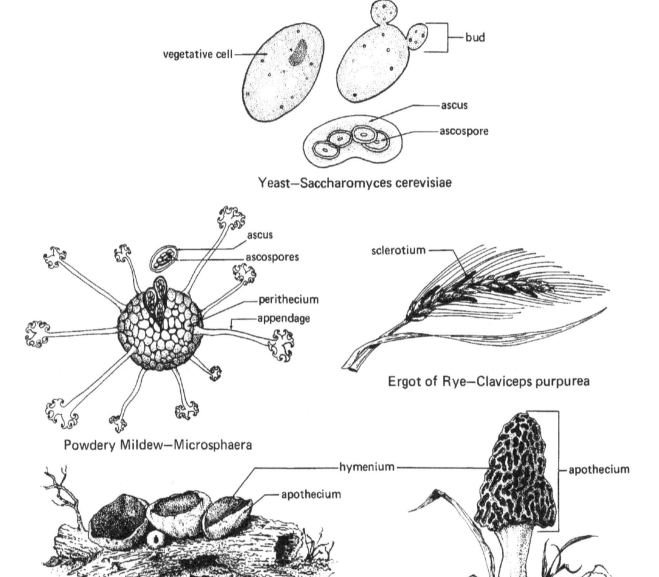

Yeast—Saccharomyces cerevisiae

Powdery Mildew—Microsphaera

Ergot of Rye—Claviceps purpurea

Cup Fungus—Peziza

Morel (Edible Mushroom)—Morchella

FIGURE 36–4

Yeasts

1. Prepare a wet mount slide from a stick culture of *Saccharomyces*. Add a drop of either cotton blue or methylene blue stain.

2. Draw individual yeast cells below:

Saccharomyces

3. Obtain the fungi *Aspergillus* and *Penicillium* that have been grown in a Petri dish. DO NOT open the sealed dish.

4. Using a dissecting microscope, observe the hyphal filaments and mycelium.

5. List the structures that you can identify.

_____

_____

IV. PHYLUM BASIDIOMYCOTA

The **Basidiomycota** is a diverse group that includes the mushrooms, shelf fungi, puffballs, and parasitic forms like the rusts and smuts (Figure 36–5). The common name, club fungi, is derived from the shape of the **basidium,** which bears the sexual **basidiospores.** Their asexual spores are sometimes **conidiospores.**

Mushrooms are the most familiar fungi (Figure 36–6). The body of the mushroom is divided into the cap (**pileus**) which is situated on a stalk (**stipe**). On the undersurface of the cap are the gills where the spores can be found. When the mushroom is young, a covering (**veil**) extends from the cap to the stalk. In the mature mushroom, the remains of the veil can be seen as ring-like **annulus.**

Commercial Mushroom
Agaricus Campestris

Fly Mushroom
(Amanita Muscaria)

Tricholoma sp.

Boletus sp.

Polyporous Cinnabarinus

Fomes sp.

Hydnum sp.

Coral Fungus
Clavaria sp.

Puffballs (Lycoperdon sp.)

Birds Nest Fungus (Lyanthus sp.)

Earth Star (Geaster sp.)

Stinkhorn
(Phallus sp.)

FIGURE 36–5

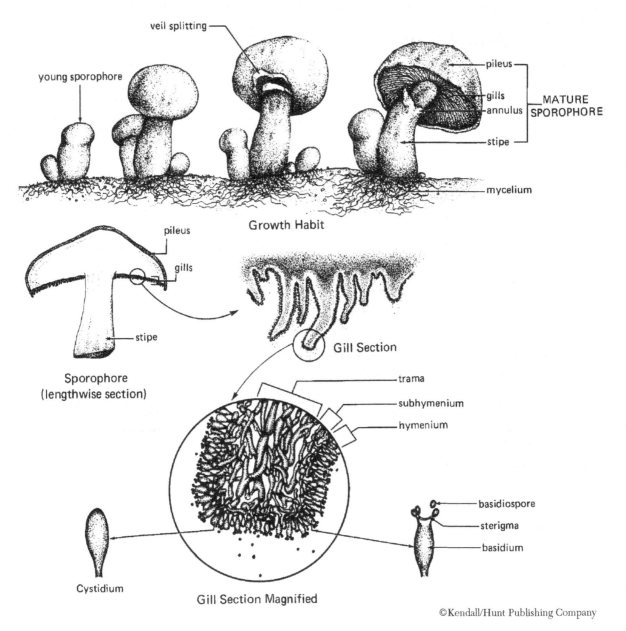

Growth Habit

Sporophore
(lengthwise section)

Gill Section

Gill Section Magnified

Cystidium

FIGURE 36–6

Procedure:

1. Obtain a fresh mushroom, *Agaricus campestris* (the white mushroom).

2. Draw the external features, and label the structures of the mushroom.

Agaricus campestris

3. Remove the cap, and with a sharp razor blade, or scalpel, cut a THIN section of the gill. (If you need help with this, call over the instructor).

4. Prepare a wet mount slide.

5. You should be able to observe a continuous **hymenial layer** along the entire gill margin. Projecting from the **hymenium** are the **basidia** that contain the **basidiospores.** **A sterigma** connects the spores to the basidium.

6. Draw this below and label:

Section of the gill

V. PHYLUM DEUTEROMYCOTA

The **deuteromycota (Fungi Imperfecti)** are the fungi that have no sexual stage. Many reproduce by the formation of **conidia** that are arranged in chains at the end of a **conidiophore** (Figure 36–7).

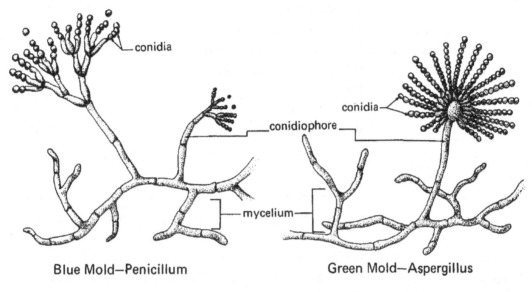

Blue Mold—Penicillum   Green Mold—Aspergillus

©Kendall/Hunt Publishing Company

FIGURE 36–7

Procedure:

1. Obtain a prepared slide of Penicillium and of Aspergillus.
2. Observe the formation of conidia.
3. Draw each below, labeling the hyphae, mycelium, conidia, and conidiophores.

Penicillium   *Aspergillus*

VI. QUESTIONS

1. _____ What is the nutritive of most fungi?

2. _____ To which kingdom do fungi belong?

3. _____ What is the name given to organisms which feed on dead or decaying
organic matter?

4. _____ Are any fungi parasites?

5. _____ What is the basic reproductive structure in fungi?

6. _____ Which hypha function similar to the "runners" of strawberries?

7. _____ What are the rootlike structures of fungi?

8. _____ What kind of spores are produced by the Phylum Ascomycetes?

9. _____ What kind of spores are produced by the Phylum Basidiomycetes?

*10. _____ Since fungi cannot make their own food, they are
_____

*11. _____ and their mode of nutrition is by _____

*12. _____ The basic filaments of fungal growth are called
_____, and _____

*13. _____ a mass of these filaments is called a _____

14. When you leave the top off of a jar of food, or leave bread uncovered, it is likely to become "moldy."
Where might the mold have come from?

_____

_____

_____

_____

_____

15. Many people are "allergic to fungi." They cough and sneeze and sniffle. Couldn't they avoid their
suffering if they just would not eat mushrooms? Explain.

_____

_____

_____

_____

_____

_____

From *Experiencing Biology: A Laboratory Manual for Introductory Biology*, 5th Edition. Copyright © 2000 by GRCC Biology, 101 Staff.
Reprinted by permission of Kendall/Hunt Publishing Company.

16. During April each year, many people come to Michigan to search for the delicious mushroom called the morel. The best mushroom-collecting season is generally during the damp periods that occur in spring and autumn. Why do you suppose that is?

_____

_____

_____

_____

_____

17. Many fungi produce antibiotics, such as penicillin, which are valuable in medicine. Of what value might the antibiotics be to the fungi that produce them?

_____

_____

_____

_____

_____

Exercise 37 KINGDOM PROTISTA—PROTOZOA AND SLIME MOLDS

OBJECTIVES

After completing this exercise, you should be able to:

1. Describe the characteristics specific to the protozoa and slime molds.
2. Identify selected example of the protozoa and slime molds according to their mode of locomotion, nutrition, cellular organization, and means of reproduction.

I. INTRODUCTION

Protozoans are one-celled, eukaryotic organisms. They are predominantly **heterotrophs,** feeding, on bacteria and small particulate matter, and inhabit water and soil. Some are **parasitic,** living, on or in other organisms. Each phylum of protozoa is classified on the basis of its mode of locomotion.

In general, protozoans live in an area that has a large supply of water. Water can be accumulated or expelled by means of a **contractile vacuole.** Food can be transported directly across the plasma membrane. However, for species that possess a protective covering, food is ingested by means of specialized structures. Digestion takes place in **vacuoles,** and the elimination of waste occurs across the plasma membrane, or through an **anal pore.**

II. PHYLUM RHIZOPODA

The **amoebas** are found would wide, in fresh and salt water, and in soil (Figure 37–1). Some forms of amoebas are parasitic. An example is *Entamoeba histolytica,* which causes amoebic dysentery. Amoebas move and feed by the use of **pseudopods** (Greek for "false food"). Pseudopods are flowing projections of cytoplasm that extend and pull the amoeba forward, or encircle and engulf food. Amoebas reproduce asexually by a process called **fission.**

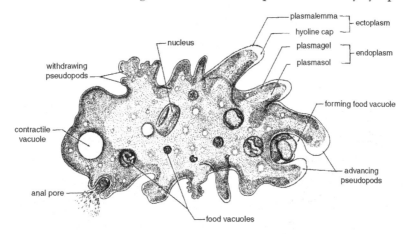

FIGURE 37–1

From *Biological Investigations* by Gayne Bablanian. Copyright © 2002 by Gayne Bablanian. Reprinted by permission of Kendall/Hunt Publishing Company.

Procedure:

1. Prepare a wet mount slide by removing a drop of water from the BOTTOM of the *Amoeba* culture using an eyedropper.
2. Under SCANNING POWER, look for granular, grayish, irregularly shaped "blobs". Then go to higher power to observe the organism.
3. Observe the *Amoeba* until you see it begin to move.
4. Describe this movement:

_____

_____

5. Add a drop of nutrient broth, and observe feeding behavior.
6. Draw the *Amoeba* below. Label the nucleus and psuedopods.

Amoeba

III. PHYLUM FORAMINIFERA

The **forams** are marine protists, and range in size from 20 micrometers to 3 centimeters (Figure 37–2). These "shelled amoebas" are surrounded by a covering (called a **test**) composed of organic materials and calcium carbonate. (The fossil remains of the tests are constituents of limestone formations, like the White Cliffs of Dover, in England.) The tests have pores through which the **podia** (feet) of the organism can emerge. Forams live in sand, are attached to other organisms, or are planktonic.

Lateral View

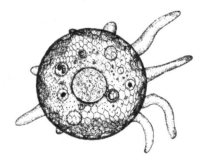

Surface View

FIGURE 37–2

Procedure:

1. Examine a prepared slide of the tests of various forms of forams.
2. Draw two different types below:

Foram

Foram

IV. PHYLUM ZOOMASTIGOPHORA

The **zoomastigotes,** or **flagellates,** are unicellular, variable in form, and have at least one flagellum (Figure 37–3). They include both free-living and parasitic types. The parasitic forms can live within various animals including humans. Some examples are *Giardia* (an intestinal parasite that causes Hiker's Diarrhea), *Trichomonas vaginalis* (a sexually transmitted protozoan), and *Trypanosoma* (causes sleeping sickness and is transmitted by the tse tse fly).

From *Biological Investigations* by Gayne Bablanian. Copyright © 2002 by Gayne Bablanian. Reprinted by permission of Kendall/Hunt Publishing Company.

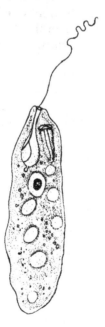

Peranema

FIGURE 37–3

Procedure:

1. Examine the prepared slides of *Trypanosoma*.
2. Note: You will also see red blood cells in this sample.
3. Draw a few blood cells and the *Trypanosoma* parasite. Label the blood cells, and the protozoan's nucleus and flagellum.

Trypanosoma

V. PHYLUM CILIOPHORA

The **ciliates** move by means of cilia, that are arranged either in spirals around the body of the organism or in longitudinal rows (Figure 37–4). The outer covering of the ciliates is called the **pellicle,** and is tough, but flexible. Because of this covering, food must be ingested through specialized structures like the **cytostome** found in *Paramecium.* The ciliates also possess two types of nuclei. The **micronuclei** are involved in reproduction and heredity, and the **macronuclei** in the control of the physiology of the cell.

A. *Paramecium*

Paramecium, a common fresh water protist, has cilia located around the cell, and in the **oral groove,** a long shallow depression which leads to the mouth or **cytostome.** Food entering the cytostome is deposited into a **food vacuole,** where the food is digested. Egestion of waste materials is through a part of the pellicle called the **anal pore.** Two **contractile vacuoles,** one located at each end of the cell, pump out excess water that has entered the cell. Asexual reproduction occurs by a process called **fission.** Sexual reproduction, if present, consists of the exchange of nuclei in a process called **conjugation.**

Procedure:

1. Prepare a wet mount slide of living *Paramecium,* using a drop of methylcellulose to slow down the organisms.
2. How does the movement of *Paramecium* compare to that of *Amoeba*?

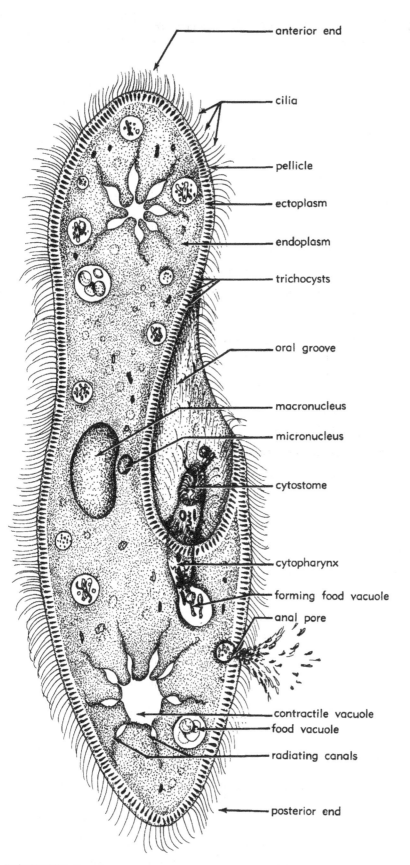

anterior end

cilia

pellicle

ectoplasm

endoplasm

trichocysts

oral groove

macronucleus

micronucleus

cytostome

cytopharynx

forming food vacuole

anal pore

contractile vacuole

food vacuole

radiating canals

posterior end

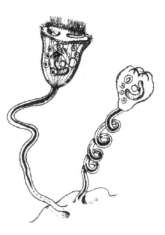

Figure E Vorticella

FIGURE 37-4

3. Draw *Paramecium* below. Using **Figure 37–4** as a guide, label as many structures as you can.

Paramecium

B. **Vorticella**

Vorticella is not free-swimming, and consists of a **bell shaped body** attached to a long **contractile stalk.** The rim of the bell contains cilia arranged in three whorls. The cilia surround the oral groove that leads into the mouth. (**See Figure 37–4,** Insert E)

Procedure:

1. Prepare a wet mount slide if living material is available or observe *Vorticella* using a prepared slide.

2. Draw *Vorticella* below. Label the cilia, body, and contractile stalk.

Vorticella

3. In the living specimen, what is the function of the moving cilia that surround the mouth?

VI. PHYLUM SPOROZOA

The **sporozoans** are non-motile, spore-bearing parasites of animals. Their spores are small, infective bodies that are transmitted from host to host. The life cycle of these organisms is complex, and involves both asexual and sexual phases. The best known among the sporozoans is the parasite that causes **malaria,** *Plasmodium* (**Figure 37–5**).

Plasmodium is transmitted from human to human by means of mosquitoes of the genus *Anopheles*. These parasites first infect the cells in the liver. In the next stage, they enter the blood stream, and infect and rupture the red blood cells, causing toxic substances to trigger cycles of fever chills.

Life cycle of *Plasmodium vivax,* the apicomplexan that causes malaria

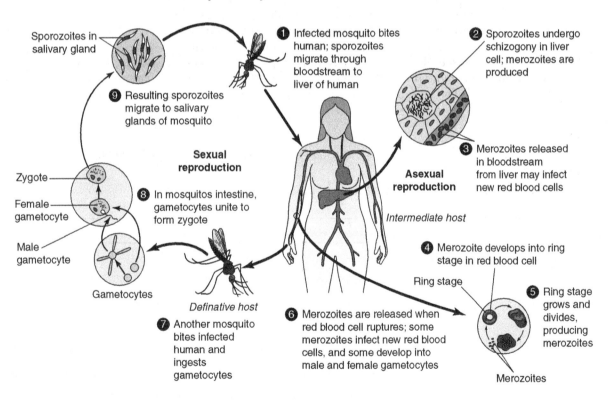

FIGURE 37–5

Procedure:

1. Examine a slide of blood with parasites inside the red blood cells.
2. Draw an infected red blood cell along with several healthy red blood cells.

Plasmodium infected red blood cells.

VII. SLIME MOLDS

Slime molds are divided into three groups, each with unique characteristics. The phylum **Acrasiomycota,** also known as the **cellular slime molds,** resemble amoebas. The phylum **Oomycota,** or the **water molds,** resemble fungi but possess cellulose in the cell wall, and have a different reproductive strategy. The phylum **Myxomycota** or **plasmodial slime molds,** exist as a mass of protoplasm with many nuclei. This mass of protoplasm is called a **plasmodium.** The entire plasmodium moves as a giant amoeba. Within the protoplasm, **cytoplasmic streaming** can be observed. The protoplasm moves so that oxygen and nutrients are distributed evenly.

Procedure:

1. Observe a demonstration plate of *Physarium,* a plasmodial slime mold. You will need to use a dissecting microscope.
2. In which direction does cytoplasmic streaming occur?

_____

_____

VIII. QUESTIONS

1. _____ Do all protists have a cell membrane?

2. _____ May protists be enclosed in a cell wall?

3. _____ How many nuclei are found typically in *Paramecium*?

4. _____ What is the structure of motility in *Paramecium*?

5. _____ What is the structure of motility in the *Amoeba*?

6. _____ What semirigid structure gives many protists their nearly constant shape?

7. _____ Protozoa are classified into groups according to their mode of
   _____ .

8. _____ *Paramecium* moves by means of its _____ .

9. _____ *Amoeba* moves by forming_____ .

10. _____ *Plasmodium* causes the disease called _____ .

11. _____ *Trypanosoma* causes the disease called _____ .

Exercise 38   PARASITIC WORM INFECTIONS OF HUMANS

INTRODUCTION

Two phyla of worms are of medical significance because they contain genera that are parasitic in humans. Members of the phylum Platyhelminthes, or **flatworms,** cause **fluke** (class Trematoda) and **tapeworm** (class Cestoda) infestations. **Roundworms,** which cause a variety of infestations, are placed in the phylum Nematoda. The two phyla of worms are commonly called **helminths,** and the science that studies them is helminthology. Helminths are studied in microbiology because diagnosis of parasitic infections in the clinical laboratory is usually by microscopic examination of body fluids, stool samples, or tissues, for the ova or larvae of the parasitic worm.

In this lab exercise, you will examine several types of parasitic helminthes, and learn to distinguish between the different classes that cause disease worldwide. For more information, see Appendix D.

A. **Platyhelminthes**

Trematodes:

Fluke infestations are diagnosed by the characteristic ova in the feces of the host. We will examine the Chinese liver fluke, *Clonorchis sinensis*, which inhabits the bile ducts, gallbladder, and pancreatic ducts of humans.

Procedure:
1. Examine a prepared slide of the adult *Clonorchis* under your scanning lens. Compare your slide to **Figure 38–1.**
2. Draw and label *Clonorchis* below.

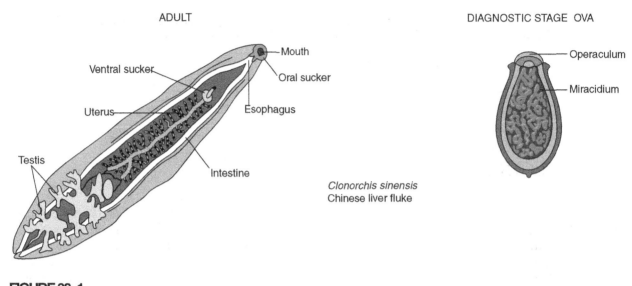

ADULT

DIAGNOSTIC STAGE OVA

Clonorchis sinensis
Chinese liver fluke

FIGURE 38–1

3. Examine a slide of *Clonorchis* ova using the high power objective.
4. Draw the ova below.

Cestodes:

Cestode, or tapeworm, infections are common in humans. Adult tapeworms live in the small intestine where they attach by a holdfast, called a scolex, which consists of suckers and hooks. The scolex is located on the head of the worm. The growing region, that is directly following the scolex, consists of segments called proglottids (see Fig. 38–2).

Procedure:

1. Examine a slide of *Taenia saginata*, the beef tapeworm, with the scanning lens.
2. Draw and label the tapeworm below.

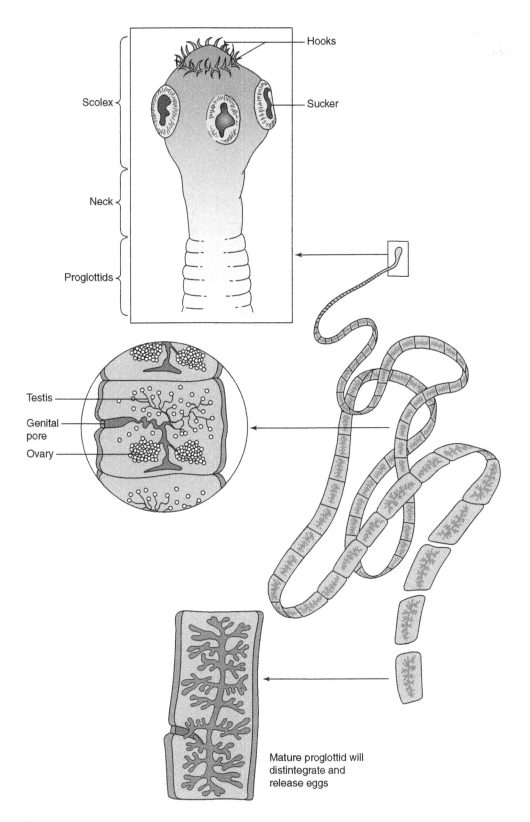

Mature proglottid will distintegrate and release eggs

FIGURE 38–2
Adult tapeworm

3. Examine preserved specimens of *T. saginata*.

4. Examine a slide of *T. saginata* ova under your high power objective.

5. Draw the ova below.

B. Nematodes

The phylum Nematoda, or parasitic roundworms, are cylindrical in shape, have unsegmented bodies, and are tapered at each end. The cuticle, covering the body, protects the worm from the gastric secretions and enzymes of digestion in the intestinal tract of the host. The ova of various species are different, and therefore have diagnostic significance. The only exception is trichinosis, where the parasite is found in tissues, and must be diagnosed by muscle biopsy and serological tests.

1. *Enterobius vermicularis,* commonly called the pinworm, is shaped like a small, straight pin. Adult pinworms are found in the large intestine. From there, the female migrates to the anus and deposits eggs on the perianal skin. (See Fig. 38–3).

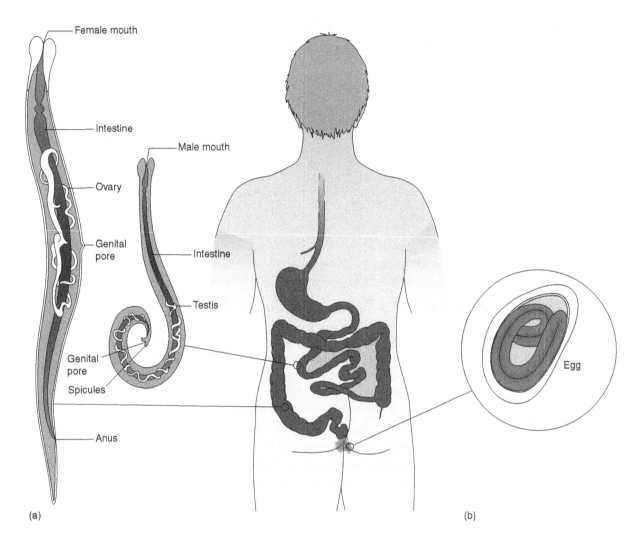

FIGURE 38–3
Enterobius vermicularis

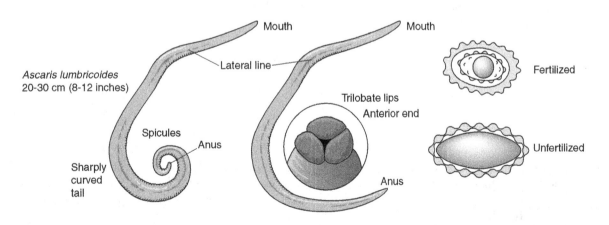

Ascaris lumbricoides
20-30 cm (8-12 inches)

FIGURE 38–4

Procedure:

    a. Examine a prepared slide of *Enterobius vermicularis* and draw below.

    b. Examine a slide of the ova of *E. vermicularis,* and draw them below.

2. *Ascaris lumbricoides* is a large nematode (30cm in length). The adult lives in the small intestine of humans and domestic animals such as pigs and horses. It feeds primarily on semidigested food. Eggs, that are excreted with the feces, can survive in the soil for long periods of time, until they are accidentally ingested by another host. See **Fig. 38–4.**

Procedure:

    a. Obtain and examine a preserved adult ascarid worm.

    b. Draw the adult worm below.

    c. Examine a slide of the ova of *Ascaris* and sketch them below.

3. The hookworm, *Necator americanus,* lives in the small intestine of humans. The hooks located around the mouth are used for attaching to and removing food from the host. Eggs, excreted in the feces, give rise to free-living larvae that inhabit soil. The larvae enter the host by penetrating the skin. See **Fig. 38–5.**

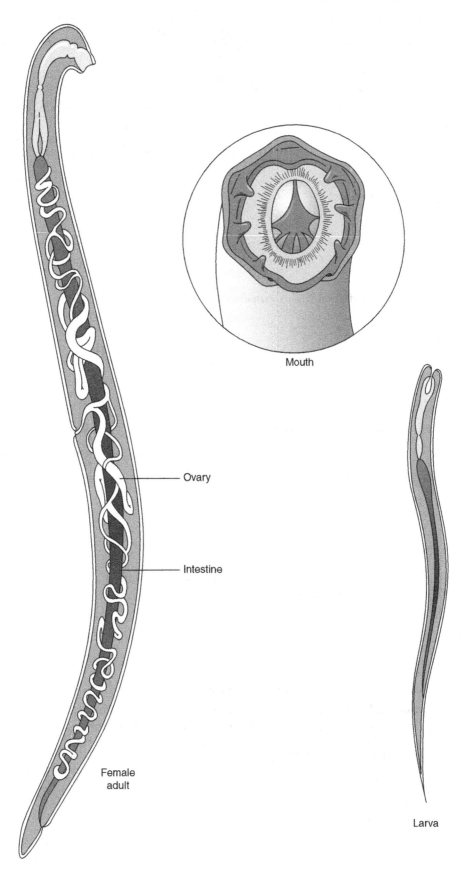

Mouth

Ovary

Intestine

Female
adult

Larva

FIGURE 38–5
Necator americanus

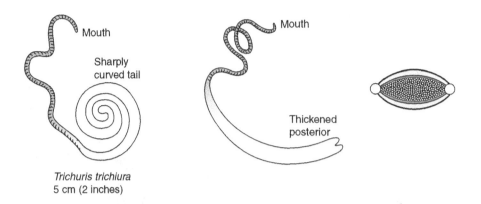

Trichuris trichiura
5 cm (2 inches)

FIGURE 38–6

Procedure:

a. Examine a prepared slide of *Necator americanus* and sketch it below.

b. Examine a slide of *Necator* ova, and sketch them below.

4. *Trichuris trichiura,* or the whipworm, is shaped like a whip. Adults attach to the lining of the cecum. Female worms produce about 5000 fertile ova per day, which are discharged in the feces. See **Fig. 38–6.**

Procedure:

a. Examine a slide of the whipworm, and draw it below.

b. Select a prepared slide of whipworm ova, and sketch them below.

5. *Trichinella spiralis* infections, called trichinosis, are usually acquired by eating encysted larvae in poorly cooked pork or bear meat. Larvae pass to the intestines, where they develop into adults. Females give birth to live nematodes, in the larval stage, that are carried through the bloodstream until they become encysted in striated muscle. See **Fig. 38–7.**

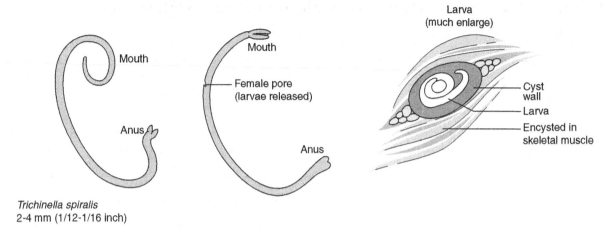

Trichinella spiralis
2-4 mm (1/12-1/16 inch)

FIGURE 38-7

Procedure:

a. Examine a slide of *Trichinella spiralis* adults and draw them below.

b. Examine a slide of larvae encysted in striated muscle tissue, and draw below.

Questions: Use Appendix D and your textbook to complete this section.

1. List two clinical symptoms of Chinese liver fluke infestation.

_____

_____

_____

_____

2. Describe the disease caused by _T. saginata._

_____

_____

_____

_____

3. List two major symptoms of pinworm infection.

_____

_____

_____

_____

4. What is the principle means of control of hookworm and whipworm infestations?

_____

_____

_____

_____

5. List three symptoms of trichinosis.

_____

_____

_____

_____

6. List three major means of control of trichinosis.

_____

_____

_____

_____

APPENDICES

LABORATORY ORIENTATION

USE OF THE PIPETTE: USE A PIPETTE PUMP WITH ALL PIPETTES

The pipette is used to transfer and measure small qualities of liquid. There are two types of pipettes in general use: the volumetric pipette and the serological pipette.

Examine the liquid in a pipette. Please note that the liquid does not appear as a straight line in the pipette. A "belly" is seen in the liquid. Water molecules are held together by **COHESIVE** forces. These forces attract similar molecules as one molecule of water to another. Water molecule are also attracted to the sides of the glass tube. These forces that attract water to the sides of the glass tube are called **ADHESIVE** forces. When the adhesive forces are greater than the cohesive forces the water molecules will tend to be pulled up the sides of a thin tube forming a belly in the liquid. This belly is called a **MENISCUS.** With most liquids we read the **BOTTOM** of the meniscus.

From *Principles of Biology: Biology 109–110, Seventh Edition* by R. Ragonese. Copyright © 2000 by R. Ragonese. Reprinted by permission of Kendall/Hunt Publishing Company.

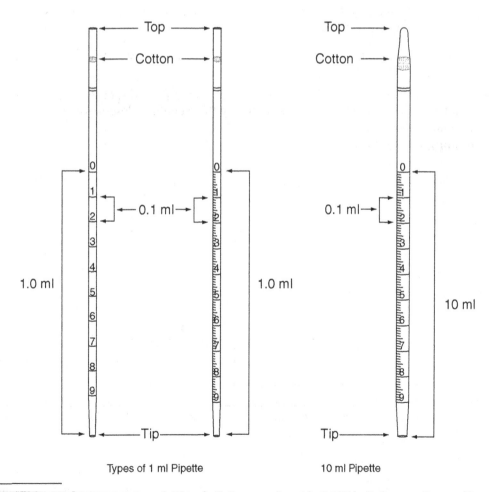

Types of 1 ml Pipette   10 ml Pipette

From *Principles of Biology: Biology 109–110, Seventh Edition* by R. Ragonese. Copyright © 2000 by R. Ragonese. Reprinted by permission of Kendall/Hunt Publishing Company.

1. The pipette will usually come in a sealed package individually wrapped or in groups of ten or twenty. Open the package just at the very top where **"PEEL HERE TO OPEN"** is clearly written.

   If you are using the pipette in an aseptic transfer operation do **not** remove the pipette from its sterile package except **immediately prior to use.** Only handle the pipette from the **top** and do **NOT** contaminate the **tip.**

2. In this case we are not concerned with aseptic transfer so just take the pipette from the package and examine it. Unless instructed otherwise assume that we are not going to use the pipette for aseptic transfer but always open the package from the very top and never at the tip end.

3. **Examine the numbers at the pipette top. You will see 10 ml in 1/10 or 1 ml in 1/10 or 1 ml in 1/100. In all cases the FIRST NUMBER is the TOTAL CAPACITY of the pipette in milliliters. The FOLLOWING NUMBERS indicate units of gradation or the unit between TWO SUCCESSIVE marked lines on the pipette side.**
   Fill in the Chart Below

Pipette Marking	Total Volume (ml)	Volume between Two Successive Lines
10 ml in 1/10		
1 ml in 1/10		
1 ml in 1/100		

4. Examine the tip. If your pipette holds 10 ml, and the number near the tip (the last number) is 9, then you have a serological pipette. If the number is 10 you have a volumetric pipette. The serological pipette holds a full capacity from zero to tip (you must release all the liquid to get full delivery of 10 ml). The volumetric pipette will hold full capacity from zero to the 10 mark. (You do not have to release all the liquid out in the tip).

5. Place five test tubes in a rack, get a 10 ml pipette, a green pipette dispenser and a 250 ml beaker. Fill the 250 ml beaker with 100 ml's of tap water. Using the technique described, pipette the following volumes into the five test tubes.

6. **The Pipetting Technique**

Test Tube	Volume
1	10 ml
2	8 ml
3	6 ml
4	4 ml
5	2 ml

From *Principles of Biology: Biology 109–110, Seventh Edition* by R. Ragonese. Copyright © 2000 by R. Ragonese. Reprinted by permission of Kendall/Hunt Publishing Company.

a. Remove the pipette from the bag. Always leave the pipette in the bag until you are ready to use it.

b. Remove the pipette by handling the top of the pipette only. Do not touch the pipette tip.

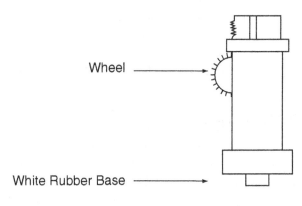

Wheel

White Rubber Base

Plastic Pipette Pump

c. Place a plastic pipette pump on the end of the pipette. These plastic pipette pumps are **NOT** discarded after use, they may be reused.

d. Use the plastic pump as instructed. Place the tip of the pipette in the liquid. Turn the wheel on the dispenser until the liquid rises to the **ZERO** mark.

e. Lift the pipette from the liquid; allow the excess fluid, on the tip, to drain. Transfer the pipette to your test tube and turn the wheel to release the liquid to the amount you want into the test tube.

f. Discard liquid remaining in pipette into the discard container. Water or saline can be discarded in sink.

g. Hold the pipette dispenser in on hand by the **WHITE** rubber base. With the other hand rotate the pipette gently until it is released form the dispenser. If the **WHITE** rubber base comes off, gently pull it free from the pipette and replace it back into the base of the pipette dispenser. **NEVER** throw the rubber base away. The dispenser will be useless if the base is missing.

PIPETTES THAT ARE USED TO TRANSFER BACTERIA MUST BE PLACED TIP FIRST IN DISINFECTANT (STAPHENE) AFTER USE.

QUESTIONS – THE USE OF PIPETTES

1. If you wished to use a 1 ml pipette to transfer 0.2 ml of fluid, what numbered line on this pipette should the liquid stop on? _____

2. If you wished to pipette 0.8 ml? _____

3. What volume is held between the tip and the 8 ml mark on a 10 ml pipette?

   _____

4. . . . between the tip and the 3 ml mark? _____

5. What is a meniscus? _____

6. How does a meniscus form in the pipette?

   _____

   _____

   _____

7. Do you read the top or the bottom of the meniscus? _____

8. Where should the pipette be kept until you are ready to use it? _____

9. What part of the pipette is never touched by hand?

   _____

10. How do you dispose of a pipette that has been used to transfer bacteria? _____

11. What is the correct way to open a bag of pipettes?

    _____

SPECTROPHOTOMETRY

I. INTRODUCTION

A SPECTROTOMETER, such as the Spectronic 20D+, is an instrument designed to measure the light absorbing properties of a solution. Biologists routinely use spectrophotometry to detect a chemical, such as sugar or protein, in a solution and determine its concentration.

Spectrophotometry is based on the observation that when a beam of light is focused on a test tube containing water, most of the light passes through the water. This light is referred to as TRANSMITTED LIGHT. The rest of the light is trapped by the water and is referred to as ABSORBED LIGHT.

If a solid is dissolved in the water, the resulting solution will absorb more of the light. The greater the concentration of the solution, the more light it will absorb.

Colored solutions absorb some wavelengths (colors) of visible light more than others. That's what gives them their distinctive color. For example, a yellow solution looks yellow, because the substance dissolved in the water transmits yellow light to your eye and absorbs the other colors (wavelengths) of light.

An ABSORPTION SPECTRUM is graphic representation of wavelengths (colors) of light a solution absorbs. Every substance has a unique absorption spectrum.

The Spectronic 20D+ can be set to display either the % TRANSMITTANCE or the ABSORBANCE of a solution. TRANSMITTANCE varies from 0% (none of the light passes through the solution) to 100% (all of the light passes through the solution). For most applications ABSORBANCE varies from 0 (no light is trapped by the solution) to 1.999 (a large amount of light is trapped by the solution). The solution to be studied is poured into a precision made glass tube (CUVETTE) and placed inside the sample chamber. Inside of the Spectronic 20D+ white light is passed through a diffraction grating that splits it into its component colors. The WAVELENGTH control knob operates a slit that selects a specific color of light to be focused on the electric tube that coverts light energy to electrical energy which is measured by a meter and can be displayed as either ABSORBANCE or TRANSMITTANCE.

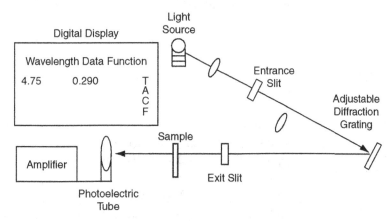

FIGURE 1
Diagram of the Path of Light through a Spectrophotometer

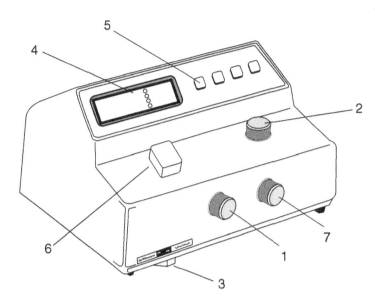

FIGURE 2
Spectronic 20D+ Spectrophotometer

KEY

1. Power Switch/Zero Control Knob (0%)
2. Wavelength Control Knob
3. Filter Level
4. Digital Readout
5. Mode Control Key
6. Sample Chamber
7. Transmittance/Absorbance Control Knob (100% T/O A)

Study Figure 2 to familiarize yourself with the control knobs and keys on the Spectronic 20D+. Then refer to it as you operate the Spectronic 20D+.

II. OBJECTIVE

Upon completion of the exercise, students should be able to:

- use the Spectronic 20D+ to measure the absorbance of a solution.
- determine the absorption spectrum of a substance.
- construct a standard curve and use it to determine the unknown concentration of a solution.

In subsequent labs, you will use the Spectronic 20D+ to determine the absorption spectrum of chlorophyll, the pigment that traps the light energy used in photosynthesis, and to follow the progress of a chemical reaction that produces a colored product.

III. PROCEDURE

A. WAVELENGTH AND COLOR OF LIGHT

1. Use the POWER SWITCH/ZERO control knob (1) to turn on the machine. The display panel will light up and a red dot will appear opposite TRANSMITTANCE indicating that the spectrophotometer is set to measure TRANSMITTANCE.

   WAIT 15 MINUTES FOR THE INSTRUMENT TO WARM UP.

2. Make sure that the sample chamber (6) is empty and its lid is closed. Then, use the POWER SWITCH/ZERO control knob (1) to adjust the display to read 0.0% TRANSMITTANCE.

3. To determine the color of a particular wavelength of light, put the paper filled into the sample chamber.

   Leave the lid open so that you can look into the sample chamber.

   a. Turn the TRANSMITTANCE/ABSORBANCE control knob (7) to the right until you can see the color of the light shining on the cuvette.
   b. Adjust the WAVELENGTH control knob (2) to set each of the wavelengths listed in Table 1 and record their colors in Table 1. Then, go on to Part B which explains how to complete Table 1.

TABLE 1
The Absorbance of a Bromophenol Blue Solution

Wavelength (nm)	475	500	525	550	575	600	625	650
Color of Light								
Absorbance								

B. ABSORPTION SPECTRUM OF BROMOPHENOL BLUE
You will measure the absorbance of a bromophenol blue solution at each of the wavelengths listed in Table 1.

1. Switch the display mode to ABSORBANCE by pressing the MODE control key (5) until the red dot is opposite ABSORBANCE. The display will flash "1999."
2. Adjust the WAVELENGTH control knob (2) until the display reads 475 nm (1 nanometer = 0.000000001 meter).

   Set the FILTER LEVER (3) to the appropriate wavelength range (340–599 nm, in this case).

3. Use a Kimwipe to remove dust and finger prints from the CUVETTE containing distilled water. (In Spectrophotometry it is referred to as the BLANK, because it does not contain the substance being analyzed.)

   The blank is used to zero the machine each time a new wavelength is selected.

4. Place the blank in the sample chamber (6) aligning the guide mark on the cuvette with the guide mark on the front of the sample chamber. Close the lid.
5. Use the TRANSMITTANCE/ABSORBANCE control knob (7) to adjust the display to read 0.0 ABSORBANCE. This may require several turns of the knob.
6. Return the blank to the rack.
7. Wipe clean the cuvette containing the 20 mg/l solution of bromophenol blue. (This solution was made by dissolving 20 milligrams of bromophenol blue in a liter of water.) Place it in the sample chamber (6) and close the lid.
8. When the digital reading stabilizes, the number displayed is the ABSORBANCE of the bromophenol blue. Record it in Table 1.
9. Return the sample cuvette to the rack.
10. Repeat steps 2–9 to determine the ABSORBANCE of the 20 mg/l sample at each of the 7 remaining wavelengths listed in Table 1.

    Be sure to switch the wavelength lever (2) to 600-950 nm when you reach a wavelength of 600 nm.

    Record your measurements in Table 1.

11. Plot the measured absorbances as a function of the wavelength on Graph 1. Notice that WAVELENGTH, the independence variable, is plotted on the X-axis, while ABSORBANCE, the dependent variable is plotted on the Y-axis. Link the data points with a series of straight lines. The resulting graph is the ABSORBANCE SPECTRUM of bromophenol blue.

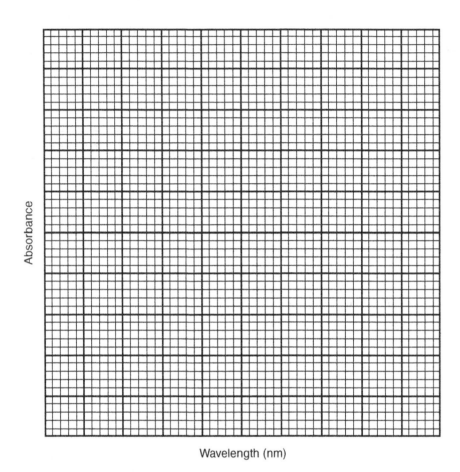

Wavelength (nm)

GRAPH 1
Absoption Spectrum of Bromophenol Blue

Which color of light was maximally absorbed by bromophenol blue? _____

Which color was least absorbed? _____

C. EFFECT OF CONCENTRATION ON ABSORBANCE

1. Adjust the Spectronic 20D+ to the WAVELENGTH at which the maximum ABSORBANCE was measured. Select the appropriate filter (lever 3). Record the WAVELENGTH in the heading of Table 11.

2. Put the BLANK into the sample chamber (6) and use the TRANSMITTANCE/ABSORBANCE control knob (7) to adjust the display to read 0.0 ABSORBANCE.

3. Remove the blank.

4. One at a time, place each of the 4 bromophenol blue solutions into the chamber (6) and read its absorbance.

5. Record each ABSORBANCE in Table 11.

6. Construct Graph 2 which displays ABSORBANCE as a function of CONCENTRATION. Draw the single straight line that best fits all the points. This technique compensates for random measurement error.

   The graph is entitled "Standard Curve for Bromophenol Blue Analysis," because it can be used to determine the concentration of a bromophenol blue solution of unknown concentration.

   Use Graph 2 to determine the concentration of a bromophenol blue solution with an absorbance of 1,000. Record the concentration below and explain how you determined it.

   _____

   _____

   _____

   _____

D. Clean-up

1. If you are the last group, turn off the Spectronic 20D+ by turning the POWER SWITCH/ZERO control knob counterclockwise until it clicks off and the display panel goes dark.

   If another group will follow you, use the MODE control key to return to the TRANSMITTANCE mode (red dot opposite TRANSMITTANCE).

2. Be sure that all 6 cuvettes have been returned to the rack. Do not empty the cuvettes.

TABLE 2

The Effect of Concentration on Absorbance of a Bromophenol Blue solution at _____ nm

Concentration (mg/l)	20	15	10	5
Absorbance				

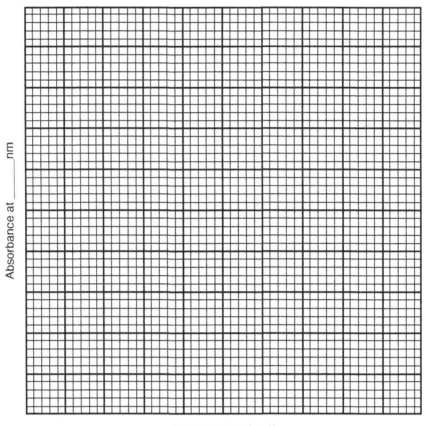

Absorbance at _____ nm

Concentration (mg/l)

GRAPH 2
Standard Curve for Bromophenol Blue Analysis

IV. REVIEW QUESTIONS: SPECTROPHOTOMETRY

1. What characteristic of a solution is measured using a spectrophotometer?

   _____

   What is the blank used for in spectrophotometry?

   _____

   What is the difference between absorbance and transmittance?

   _____

2. What was the purpose of placing a paper-filled covette in the sample chamber of the spectrophotometer?

   _____

   _____

3. What gives a solution its distinctive color?

   _____

   What is the absorption spectrum of a solution?

   _____

   _____

4. How does changing the concentration of a solution affect its absorbance?

   _____

5. For what purposes do biologists use spectrophotometers?

   _____

   _____

   _____

   _____

FURTHER STUDIES

I. MAKING WINE

Your instructor will demonstrate alcoholic fermentation by yeast by setting up from a babblenman flask containing the ingredients needed to make wine. The procedure below is only a demonstration, and should not be used for home wine making because it is not set up under sterile conditions.

PROCEDURE

1. Combine into a large flask the following ingredients:
   a. 1 liter of boiled water
   b. 1 cup of sugar
   c. 2 cups of raisins
   d. 1 package of yeast
2. Swirl to mix.
3. Cover top of flask with parafilm. Poke some holes in the parafilm to allow carbon dioxide to escape.
4. After 1 week, remove the parafilm, and smell the wine. What does it smell like?

_____

5. If the wine has been contaminated by bacteria that produce acetic acid, what will happen to the wine?

_____

II. PREPARATION OF SAUERKRAUT

Many years ago, before the days of refrigeration, the preservation of food products was a major concern. Many methods of preservation were developed, and they frequently involved the addition of salt, sometimes with other ingredients, under defined conditions. Various meat products, such as corned beef and pastrami, were developed in this way. Pickles were developed as a means of preserving a seasonal vegetable (the cucumber). Sauerkraut is a product that was developed as a method of preserving cabbage by allowing it to ferment naturally by its indigenous microbial flora. Upon completion of the fermentation, sauerkraut contains at least 1½% acid of which the majority is lactic acid. It is prepared by fermenting shredded or chopped cabbage in the presence of 2–3% salt. The cabbage and salt are mixed in a container with no exposed metal (a crock or plastic jar work well). The osmotic pressure induced by the salt causes water to flow out of the cabbage, thus producing a brine. A heavy weight (e.g., a jug of water or a rock wrapped in plastic) placed on top of the cabbage keeps it submerged in the brine and allows the fermentation to proceed in the absence of oxygen. No inoculum is necessary because the required bacteria occur naturally on the cabbage.

From A _Laboratory Manual for Microbiology_ by J. M. Larkin. Published by Kendall/Hunt Publishing Company.

The fermentation has three steps, each associated with a particular bacterial flora. At first, coliforms ferment the sugars and produce acids which lower the pH. Within a few days the coliforms give way to *Leuconostoc mesenteroides* which will produce from 0.7% – 1.0% lactic acid as well as other end products. A little later *Lactobacillus plantarum* will predominate and it will increase the total acidity to 1.5%–2.0%. If *L. brevis* is present the acidity may become even greater, resulting in an undesirable sharp flavor.

The high acidity produced during the fermentation prevents the growth of undesirable bacteria, while yeasts and molds are retarded by the anaerobic conditions.

The fermentation is generally completed in about 3 weeks, but a better flavor is usually achieved with a 4–5 week incubation period.

MATERIALS

One hand of cabbage for every 3 students

Balance

Beakers, about 1 liter size, to weigh the cabbage

Salt

Sharp knives or slicers

Pipette jar

Jug of water for weight (a 5 lb acid bottle fits a pipette jar well)

Plastic wrap

PROCEDURE

FIRST DAY

1. Remove the outer leaves of the cabbage and cut the head in half.
2. Remove the core and any bruised tissue and then wash the cabbage.
3. Shred the cabbage into long thin pieces, weigh it and then weigh out enough salt to obtain 2½% concentration (2.5g salt to 100g cabbage).
4. Layer some cabbage into the pipette jar, then add some of the salt. Continue adding layers of cabbage and salt until it is all added.
5. Add the jug of water to the top of the cabbage and place the pipette jar into the sink to catch any overflowing brine.

SECOND DAY

1. The next day, cover the top of the pipette jar and water jug with plastic wrap to keep insects out.

From *A Laboratory Manual for Microbiology* by J. M. Larkin. Published by Kendall/Hunt Publishing Company.

FOLLOWING PERIODS

1. At each of the next five periods, prepare a Gram stain from the liquid above the cabbage. Use a clean loop. After heat fixing the smear, rinse it lightly in tap water to remove the salt crystals and then proceed with the stain. Also, one person should be assigned each period to take the pH of the fluid. A description of the microbial changes that occur should be made and compared to a graph which depicts the changes in the pH.

2. When the sauerkraut is done, it may be desirable to conduct an *organoleptic* examination of the product. To do this, discard about half of the liquid and replace it with water to decrease the saltiness. Simmer the sauerkraut for about a half hour, cool it and then taste it. This examination can be made more interesting by throwing in a few hot dogs before simmering.

From *A Laboratory Manual for Microbiology* by J. M. Larkin. Published by Kendall/Hunt Publishing Company.

III. CHEESE MICROBIOLOGY

PURPOSE

To observe and isolate microorganisms from cheese.

BACKGROUND INFORMATION

Cheese is probably the most important diary product derived from microorganisms. The cheese used in this exercise are hard and semisoft cheeses which are produced in two steps (Figure 33.1). First, bacteria (streptococci) ferment the **milk sugar, lactose** to produce acid which causes the **milk protein, casein,** to gel forming **curds.** This process is speeded up by using an enzyme called **rennin** usually obtained from the stomach lining of a calf. Curd is separated from the remaining liquid, **whey.** This curd is basically **unripened** and is sold as a soft cheese, such as cottage cheese or ricotta cheese. The second step in obtaining a harder cheese is the **ripening** process in which the cheese undergoes changes in flavor, texture and consistency. Microorganisms, produce the most distinctive changes and vary according to the type of cheese desired. They are added to the curd or spread on the surface at the beginning of the ripening process. The ripening process may take weeks or months depending upon the hardness of the cheese. Cheddar cheese, which comprises about 75% of all cheese made in the United States, is produced by the addition of the organism *Lactobacillus* to the unripened curd. Blue cheese is ripened by a fungus, *Penicillium roqueforti,* which grows throughout the curd resulting in its characteristic blue-streaked appearance. The soft cheese Limburger has the principle ripening agent *Brevibacterium linens* which is inoculated on the surface of the curd. Table C-1 lists some common cheeses and their ripening agents C-2.

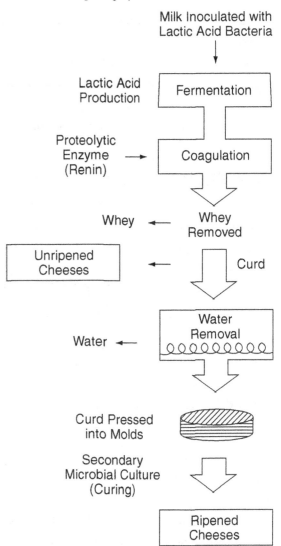

FIGURE C-1

Steps in the ripening process of cheese.

From *Microbiology in Today's World* by Barbara K. Hudson. Copyright © 1998 by Kendall/Hunt Publishing Company. Reprinted by permission.

From *Microbiology in Today's World* by Barbara K. Hudson. Copyright © 1998 by Kendall/Hunt Publishing Company. Reprinted by permission.

TABLE C-1

Cheese

Type and Name	Ripening Agent
Unripened:	
Cottage	None
Cream	None
Soft ripened:	
Camembert	*Penicillium camemberti*
Brie	*P. camemberti* and *Brevibacterium lines*
Semisoft ripened:	
Limburger	*B. linens*
Roquefort (Blue)	*Penicillium roqueforti*
Brick	*B. linens*
Muenster	*B. linens*
Hard ripened:	
Cheddar (American)	*Lactobacillus casei*
Swiss	*Propionibacterium shermanii*
Gruyere	*Propionibacterium freudenreichii*

MATERIALS

cheddar cheese

blue cheese

Limburger cheese

plates of tryptone-glucose-beef extract agar (TGE)

inoculating loop

slides

Gram's crystal violet stain

PROCEDURE

DAY 1

1. Examine the blue cheese for fine straight holes. Look for veins of blue mold. Make a smear of the mold with a drop of water and examine under hi-dry magnification.
2. Examine the Cheddar and Limburger cheeses for odor and texture. Make smears with a little cheese and a drop of water. Stain the smears with Gram's crystal violet (all of the organisms are Gram-positive), and examine under hi-dry and oil immersion. Record your observations.

3. Use an inoculating loop to streak a TGE agar plate with the Limburger cheese. Invert the plate and incubate at room temperature in daylight for at least five days. *B. linens* is a Gram-positive pleomorphic organism that needs light to produce an orange pigment.

DAY 2

1. Examine the TGE plates. If orange colonies are present, make a smear, stain with Gram's crystal violet and examine under oil immersion.
2. Indicate your observations of all the cheeses by describing and drawing their appearance.

   Results:

QUESTIONS

1. What is the difference between ripened and unripened cheese?

   _____

   _____

   _____

2. Which cheeses are ripened by molds? Which by bacteria?

   _____

   _____

   _____

   _____

3. What procedure would you follow to purify the *Brevibacterium* and obtain a culture of only this organism?

   _____

   _____

   _____

   _____

NOTES

IV. EFFECTIVENESS OF A SEWAGE TREATMENT PLANT

He who drinks a tumbler of London water has literally in his stomach more animated beings than there are men, women, and children on the face of the globe

—Sydney Smith, 1834

PURPOSE

To determine the effect a wastewater treatment plant has on the sanitary quality of river water.

BACKGROUND INFORMATION

This exercise uses the presumptive test to evaluate the effectiveness of a wastewater treatment plant. Multiple tubes of lactose broth are used to test for the presence of the gas-producing **coliform bacteria.** The coliform group of bacteria are present in our gut and feces and are used as an indication of fecal contamination of water. In other words, if they are present, intestinal pathogens may also be present. A criterion used to test for coliforms is the ability of these organisms to ferment the sugar lactose and produce gas. Gas production is detected by gas entrapment in a small inverted, liquid-filled tube inside the broth (a Durham tube). A series of lactose broth tubes is inoculated in triplicate with different dilutions of the original sample. At some dilution the coliform bacteria should be diluted to extinction, leaving no coliform cells to initiate growth and gas production. Form the pattern of positive tubes in the dilution series, a statistical estimate is made of the number of coliform bacteria in the sample (**Most Probable Number—MPN**). The MPN is an index of the number of coliform bacteria that, more probably than any other number, would give you the results shown by the laboratory analysis. In this experiment we are only performing the presumptive test which gives the MPN of *possible* coliforms bacteria. A confirmed test should also be done to determine the actual number of coliforms in each sample. Several samples of water taken from the river and various places within the treatment plant will be tested. **Figure C-3** is an abbreviated flow chart showing how sewage influent is treated before it is released into the river. The samples you will test were taken at different sites during the treatment.

MATERIALS

several actual water samples from the river collected above and below the wastewater treatment plant
untreated sewage and samples from various location within the treatment plant
tubes of lactose broth (9 ml each)
pipettes

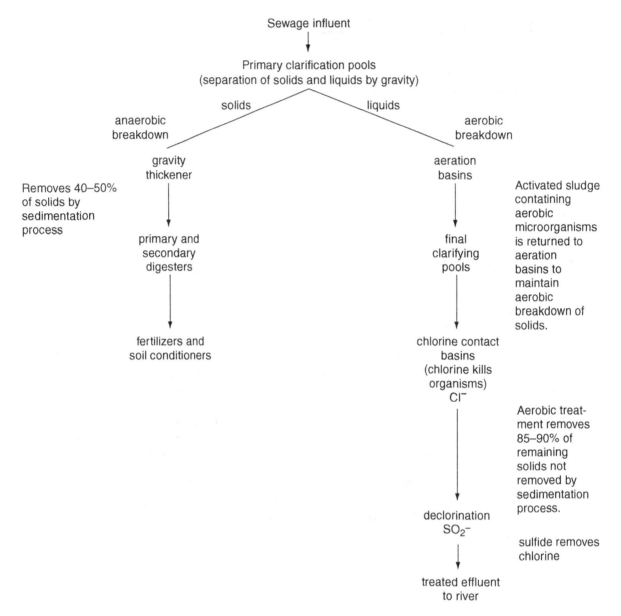

FIGURE C-3
A schematic diagram showing process used to treat sewage before it is discharged into a river.

PROCEDURE

DAY 1

Each lab section will perform an MPN analysis on untreated sewage and on one of the other samples. A six seat by six row pattern will be used to set up a triplicate dilution series for each sample (the sixth seat dilution will be performed by the TA). Rows A, B, and C will dilute one sample and rows D, E, and F the other. Each sample is diluted serially down the row from 1:10 (10^{-1}) to 1:000000 (10^{-6}) as follows:

1. At seat one in each row, use a 1.0 ml pipette to aseptically transfer 1 ml of the sample to a tube containing 9 ml of lactose broth, making a 1:10 or a 10^{-1} dilution. Mix the tube well but *do not* invert the tube (caps are not liquid-tight) and *do not* introduce air into the inverted tube. Label this tube 10^{-1}. Discard the pipette into the pipette discard.

2. At seat two, use a fresh sterile pipette to aseptically transfer 1 ml of the 1:10 dilution into a second 9 ml tube to lactose broth, making a 1:100 dilution. Mix it well and label this tube 10^{-2}. Discard the pipette.

3. At seat three, similarly transfer 1 ml of dilution 10^{-2} to a tube of 9 ml of lactose broth, making a 1:1000 dilution. Mix it well and label this tube 10^{-3}. Discard the pipette.

4. Likewise, dilute serially at each seat down the row until a 10^{-6} dilution is reached at seat six. Label each tube with its dilution. Below is an illustration of the dilution process of the six tubes in each row:

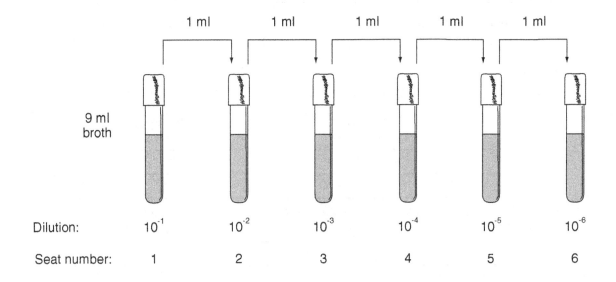

5. Put the tubes from your row in a small container. Put a paper in the container stating: a. your laboratory hour; b. your room number; c. your row letter (A, B, C, etc.); and d. the type of water sample. Put the container in a box provided for incubation at 35°C. The tubes will be incubated until the next laboratory period.

DAY 2

1. Read all tubes with any gas trapped in the inner tube as presumptive positive.
2. Record the results for all replicates and all dilutions (+ or −) in the table below. Your TA will put results for the entire class on the board.

Row	Dilution					
	10^{-1}	10^{-2}	10^{-3}	10^{-4}	10^{-5}	10^{-6}
A						
B						
C						
D						
E						
F						

3. The TA will calculate form an MPN table or the formula given at the end of this exercise the most probable number of coliform bacteria in your sample. Record your results below. Observe the class results for all samples. The United States Environmental Protection Agency has established standards based upon the maximum number of coliform organisms allowable per 100 ml of finished water. An MPN of less than 2.2 coliforms/100 ml is generally interpreted to indicate that the sample meets the standards.

Sample _____

MPN of Coliform Bacteria _____

QUESTIONS

1. How many coliform bacteria were present in the untreated sewage? In the water sample?

_____

_____

_____

2. How effective is the sewage treatment plant in improving the sanitary quality of untreated sewage?

_____

_____

_____

3. Does the plant influence the bacteriological quality of the river?

_____

_____

_____

4. What steps in waste treatment cause the most significant improvement in water quality?

_____

_____

_____

_____

INSTRUCTOR NOTES

The MPN can be determined from a table giving an MPN index such as the one from the 16th edition of *Standard Methods for the Examination of Water and Wastewater,* American Public Health Association, 1985 or can be calculated using the following formula:

$$\text{MPN/100 ml} = \frac{\text{Number of positive tubes} \times 100}{(\text{ml sample in negative tubes}) \times (\text{ml sample in all tubes})}$$

NOTES

TABLE 1
MPN Determination From Multiple Tube Test—Appendix A

3 OF 10 ml each	3 OF 1 ml each	3 OF 0.1 ml each	MPN INDEX PER 100ml
0	0	1	3
0	1	0	3
1	0	0	4
1	0	1	7
1	1	0	7
1	1	1	11
1	2	0	11
2	0	0	9
2	0	1	14
2	1	0	15
2	1	1	20
2	2	0	21
2	2	1	28
3	0	0	23
3	0	1	39
3	0	2	64
3	1	0	43
3	1	1	75
3	1	2	120
3	2	0	93
3	2	1	150
3	2	2	210
3	3	0	240
3	3	1	460
3	3	2	1,100

V. PETROLEUM USE BY BACTERIA

PURPOSE

To observe the growth of bacteria on oil.

BACKGROUND INFORMATION

Petroleum is a rich source of organic matter. It is not surprising, therefore, that a wide variety of microorganisms will readily attack it under certain environmental conditions. This demonstrates a beneficial role of microorganisms—the **biodegradation** of pollutants. Significant breakdown will only take place in the presence of oxygen. If the oil gets carried into an anaerobic environment (absence of oxygen), it will not be decomposed and may remain in place for many years. This helps to explain why natural oil deposits may be millions of years old. On the other hand, because oil is insoluble in water and is less dense, it will float on the surface of water and form oil slicks. The oil is exposed to oxygen and is quickly attacked by oil-degrading bacteria which eventually decompose the oil the disperse it. Petroleum degraders include Gram-positive bacteria, often with irregular shapes, Gram-negative bacteria and certain molds and yeasts. Most soil should have some bacteria which are capable of using oil. Service stations usually have oil-soaked soil which is an excellent source of these microorganisms.

> Many researchers are investigating microbial mechanisms for degradation of compounds. Several at Montana State University are using fungi to bioremediate soil and water contaminated with heavy metals. While some metals, such as iron and copper, are essential for biological growth, heavy metal accumulation is toxic to most organisms, including humans. Scientists have found that pigmented fungi bind heavy metals more readily than non-pigmented fungi and are investigating ways of using hyphae and synthetic pigments to bind metals from polluted, abandoned mining sites.

MATERIALS

    a soil sample, oil-soaked, if possible
    a tube of mineral medium with mineral oil added as the only carbon source
    one Erlenmeyer flask of the same medium for each class
    slides
    cover slips
    Gram stain reagents
    staining racks

PROCEDURE

DAY 1

1. Add a small amount of soil to the tube of mineral oil medium and to the Erlenmeyer flask.
2. Incubate the tube slanted to increase surface area in your locker. The Erlenmeyer flask will be incubated on a shaker and should have growth in one week. The tube may take longer.

DAY 2

1. When the flask and tube have become turbid, make a smear. Heat fix and Gram stain it. Observe the organisms present.
2. Record your results below.

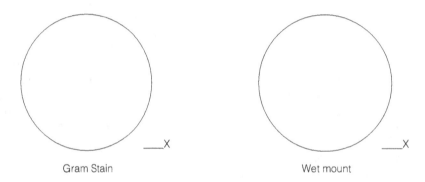

Gram Stain   Wet mount

QUESTIONS

1. Why was the Erlenmeyer flask incubated on a shaker?

   _____

   _____

   _____

   _____

2. What other environmental factors besides oxygen might influence the rate of oil degradation?

   _____

   _____

   _____

   _____

3. What special requirement do oil-degrading microorganisms have in order to obtain their energy source?

_____

_____

_____

_____

NOTES

VI. CONTAMINATION OF A DRINKING GLASS

PURPOSE

To demonstrate the transfer of bacteria from the mouth to a drinking glass. To demonstrate the use of a diagnostic medium to study the prevalence of streptococci in the mouth.

BACKGROUND INFORMATION

From previous exercise it was leaned that there are millions of microorganisms in our mouth. Therefore, it would be a logical assumption that the lips and anything they cam in contact with could also have a similar microbial flora. The bacteria would be **transferred** from the mouth to the drinking glass. As you observed from growing and staining bacteria from the mouth, one of the predominant organisms is streptococci (Gram-positive cocci in chains). They can be selected for by growth on a sucrose agar medium. Streptococci, especially *Streptococcus salivarius*, convert the sugar sucrose to a sticky dextran material and thus will grow as raised mucoid colonies, making this an excellent **diagnostic medium.** Other organisms from the oral cavity will be inhibited because they will not grow in such a high concentration of sucrose (5%).

MATERIALS

plates of sucrose agar
sterile drinking glasses
tubes of sterile water
sterile swabs
glass slides
Gram's crystal violet stain

PROCEDURE

DAY 1

1. Work in pairs for this exercise. On the bottom of the sucrose agar plate, divide the plate in two with a line from a wax pencil. Label one half of the plate "lips" and the other half "glass". Do not label the lid—only the bottom of the plate.
2. Obtain a sterile swab and moisten it in a sterile water. Remove any excess water from the swab by pressing it against the inside of the tube.

3. Swab the lips of one partner and streak the half of the plate labeled "lips". Be sure to use the same side of the swab at all times. It is not necessary to streak for isolation. Discard the swab in the proper container.

4. Carefully, remove the paper from the top of the sterile glass (avoid touching the rim of the glass). The partner whose lips were swabbed should then touch their lips to the glass in several places as if drinking from it.

5. Moisten a second sterile swab and swab the inside and outside of the rim of the glass several times. Then, streak the half of the agar plate labeled "glass". Discard the swab.

6. Invert the plate and incubate according to the TA's instructions until the next lab period.

DAY 2

1. Examine the sucrose agar plate and look for raised mucoid colonies characteristic of *Streptococcus salivarius*. Notice the profile of these colonies and draw an outline of a side view of the colony.

2. Make a smear of the colony and stain 1 minute with crystal violet. You need not do an entire Gram stain as streptococci are easily recognized as Gram-positive cocci in chains. Examine under oil immersion.

QUESTIONS

1. What shape and arrangement of cells did you see?

   _____

   _____

   _____

2. Teeth scrapings show a mixed culture in the mouth. Why did you grow just one bacterium in the sucrose agar medium?

   _____

   _____

   _____

   _____

3. Why was it important to use the same person for both parts of the experiment?

   _____

   _____

   _____

   _____

NOTES

ADDITIONAL INFORMATION ON PARASITIC INFECTIONS OF HUMANS

FUNGI

Rhizopus nigricans:

This is one of the most common saprophytic fungi known. It grows abundantly on stale bread, on almost any culture medium, and occasionally on living plants. *Rhizopus* has the following structures:

RHIZOIDS – anchor, secrete digestive enzymes, and absorb nutrients.

STOLONS – horizontal hyphae that enable the fungus to spread rapidly. Functions in asexual reproduction. The stolon functions like runners on a strawberry plant.

SPORANGIOPHORES – upright hyphae which produce dark, spherical sporangia at their tips.

SPORANGIA – structures that produce and contain spores.

SPORES – resists dryness and functions in asexual reproduction.

COLUMELLA – a supportive structure found at the base of the sporangium.

Sexual reproduction occurs when hyphae from two different fungi produce PROGAMETANGIA, which grow towards each other. Each progametangium forms a SUSPENSOR and a GAMETANGIUM (containing gametes). The two gametagia fuse to form a ZYGOTE, which develops a thick wall and becomes a ZYGOSPORE (resists dryness).

Penicillium

This fungus is very common and widely distributed in nature. It is a saprophyte on all kinds of organic matter including bread, cured meat, cotton, woolen fabrics and leather. Camembert, Roquefort and other cheeses owe their properties and flavors to species of *Penicillum*. It is also the source of the antibiotic penicillin.

Asexual reproduction is by CONIDIA formed in chains at the ends of bowling pin-shaped STERIGMATA. The conidia, which resists dryness, are produced in great numbers and are distributed by air currents.

Fungal Diseases of Humans:

1. Some people are more susceptible to fungal infections than others. Risk can be heightened by heat, moisture and suppression of the immune system.

2. Fungal infections are contagious and must be treated promptly. If untreated, complications may result and other people may become infected. Fungal skin infections can look like rashes due to other causes.

3. Scalp fungal infections are very contagious in children. Round, gray, scaly bald patches are common symptoms. Sometimes the infections may appear pink, swollen, and oozing. Scarring and permanent baldness may result. Infection may spread by common use of hats.

PARASITIC PROTOZOA

Parasitic protozoa are unicellular animals. Each cell consists of mostly cytoplasm and a nucleus. The following definitions will be helpful.

TROPHOZOITE – the motile form of the parasite which feeds and multiplies.

CYST – The immobile form protected by a cyst wall and designed for transmission to new hosts.

ENCYSTATION – The transformation of a trophozoite into a cyst.

EXCYSTATION – Hatching of the cyst with the liberation of a motile trophozoite.

SCHIZOGAMY – Asexual multiplication by fission.

SCHIZONT – A trophozoite in which the nucleus has divided or split.

MEROZOITE – One of the cells resulting from the division of a schizont.

TRYPANOSOME – A flagellate protozoan found in blood smears.

Entamoeba histolytica

The parasite is by far the most pathogenic of the intestinal protozoa. It is the cause of amebic dysentery and other less sever intestinal disturbances as well as amebic abscess of the liver and other organs. The preferred habit is the colon. The organism exists in two forms, the trophozoite and the cyst. The trophozoite is fragile and dies quickly outside the host. The cysts are hardy and designed for the purpose of transfer of new hosts. The ingested cysts undergo excystation in the lower ileum, then move to the colon to parasitize the new host.

Trypanosoma sp.

Trypanosoma are found in the plasma between the red corpuscles. Infection with trypanosomas is known as TRYPANOSOMIASIS. The important type to humans are the African Sleeping Sickness and Chagas Disease from South America. The African variety is transmitted or vectored by the tsetse fly. Examine the prepared slide of *Trypanosoma brucei*. This parasite does not attack humans, but is easy to see under the microscope. Some of the slides are *Trypanosoma rhodesiense*.

Giardia lamblia

This flagellate lives in the small intestine. It has a sucking disc, which serves for attachment to the wall of the gut. *Giardia* reproduces by longitudinal fission. It encysts probably in the ileum. Within the cyst the various organelles are duplicated, and two trophozoites are liberated with excystation.

Transmission to a new host occurs via the cysts through fecal contamination of food and water. Excystation occurs in the small intestine.

Plasmodium vivax

This is the protozoa that causes malaria. It is transmitted or vectored by the *Anopheles* mosquito. Only the female mosquito sucks blood. The males live off the nectar from flowers. *Plasmodium* has adapted itself to a life within the red corpuscles of the host.

Examine the slid of *Plasmodium vivax* in the blood smear. Remember the parasite will be found inside the red corpuscle. Identify the following stages and their features:

YOUNG TROPHOZOITE – This is often called the "ring stage" because it resembles a signet ring. Note the CYTOPLASM, NUCLEUS, and the VACUOLE.

GROWING TROPHOZOITE – The shape of the parasite is more irregular. Note the CYTOPLASM, NUCLEUS, and the VACUOLE.

MATURE SCHIZONT – This is a very important stage. Note the large number if NUCLEI present. When the red corpuscle containing the mature schizont is destroyed, each nucleus and a bit of cytoplasm is organized into a merozoite that is capable of infecting other red corpuscles.

For each of the parasitic protozoa listed above know the following information:

1. Two diagnostic symptoms of the disease produced by the parasite.
2. Source and mode of infection.
3. Laboratory diagnosis.
4. Methods of preventing the disease.

PARASITIC PROTOZOAN DISEASES

Amebiasis

A. Symptoms:
   1. Blood or mucoid diarrhea alternating with periods of constipation.
   2. Chills and fever.
   3. Liver abscesses if parasite migrates to the liver.
B. Diagnosis: Identification of ameboid protozoans in the patient's feces.
C. Sources and Mode of Infection:
   1. Ingestion of fruits and vegetables contaminated by human feces as a result of:
      a. Flies, which after walking across feces, also strolls across your food.
      b. Food handlers contaminating food.
      c. Water contaminated by sewage.
D. Prevention:
   1. Persons found to be excreting *Entamoeba histolytica* in their feces should be prohibited from food handling.
   2. A safe water supply.
   3. Strict sanitation in food handling establishments.
   4. Improve sanitation.

Trypanosomiasis

A. Symptoms:
   1. Enlargement of lymph nodes, spleen, and/or liver.
   2. Swelling of eyelids.
   3. Inflammation of the heart muscle.
   4. General debility.
   5. Unnatural drowsiness (sleeping sickness).
B. Diagnosis: Presence of trypanosomas in the blood, and in later stages in the cerebrospinal fluid (CSF).
C. Source and Mode of Infection:
   1. South American Form:
      a. Infection is via contamination of the conjunctiva, mucous membranes or skin by fecal contamination by infected vectors.
      b. Blood transfusions from infected donors.

   2. African Form:

      a. The bite of the tsetse fly or horse fly.

D. Prevention:

   1. Elimination of vectors and infected domestic animals.

   2. Wear protective clothing.

Giardiasis

A. Symptoms:

   1. Chronic diarrhea.

   2. Abdominal cramps.

   3. Bloating.

   4. Fatigue.

   5. Weight loss.

   6. Indigestion.

B. Diagnosis: Identify the cyst or trophozoites in the patient's feces.

C. Source and Modes of Infection:

   1. Surface water contaminated by animals or humans.

   2. Poor sanitation.

   3. Ingestion of contaminated food or water.

   4. Hand to mouth transfer of cysts from the feces of an infected individual.

D. Prevention:

   1. Efficient sanitation.

   2. Protection of water supplies against fecal contamination.

   3. Education of people in personal hygiene.

Malaria

A. Symptoms:

   1. Characteristic chills followed by fever.

   2. Nausea.

   3. Headaches.

   4. Body pain.

B. Diagnosis: Presence of malaria parasites in blood smears.

C. Source and Mode of Infection:

   1. Bite of an infected female *Anopheles* mosquito.

   2. Blood transfusions from an infected donor.

   3. Use of contaminated hypodermic syringes by drug addicts.

D. Prevention:

   1. Destroy the mosquito breeding places. Drain the swamps.

   2. Use mosquito fish.

   3. Prophylactic doses of an anti-malarial drug.

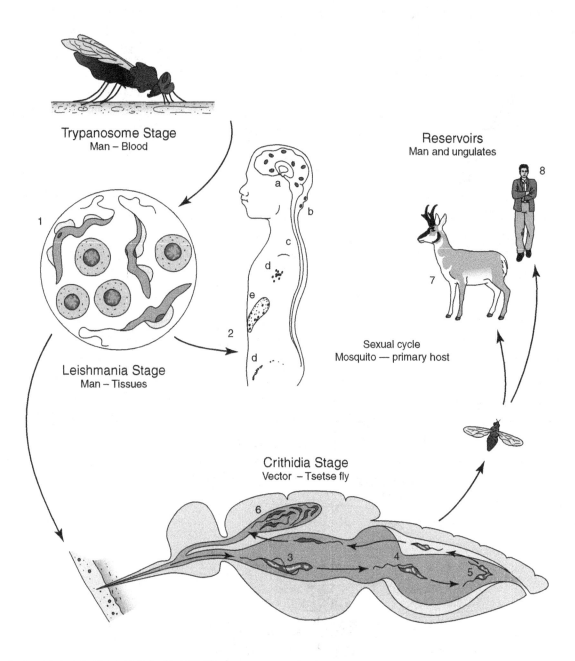

Trypanosome Stage
Man – Blood

Leishmania Stage
Man – Tissues

Reservoirs
Man and ungulates

Sexual cycle
Mosquito — primary host

Crithidia Stage
Vector – Tsetse fly

TRYPANOSOME LIFE CYCLE

1. Trypanosomes in blood
2. Trypanosomes in tissues
   a. Brain
   b. Cervical lymph nodes
   c. Spinal cord
   d. Other lymphoid tissue
   e. Spleen

3. TRYPANOSOME (long forms)
4. Crithidia
5. Crithidia (binary fission)
6. Metacyclic trypanosomes
7. Reservoir
8. New host

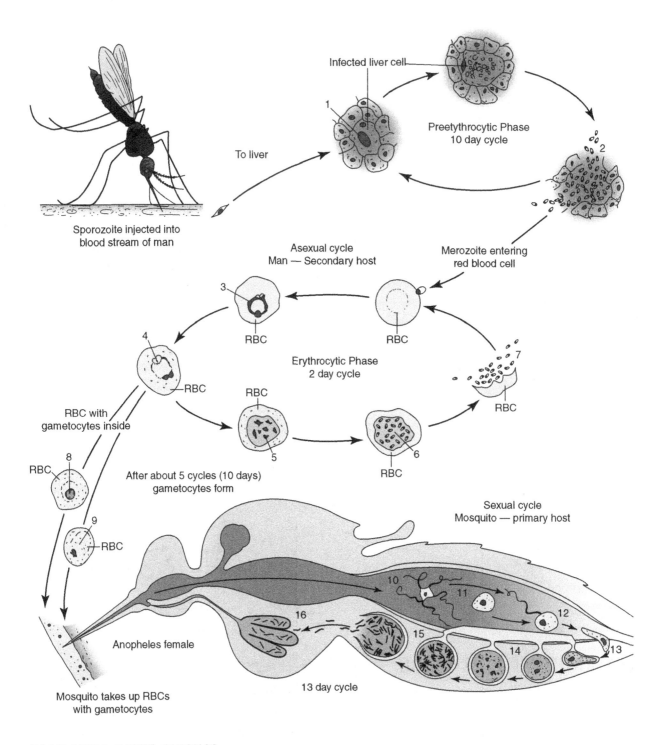

Infected liver cell

Preetythrocytic Phase
10 day cycle

To liver

Sporozoite injected into
blood stream of man

Merozoite entering
red blood cell

Asexual cycle
Man — Secondary host

RBC

RBC

Erythrocytic Phase
2 day cycle

RBC

RBC

RBC

RBC

RBC with
gametocytes inside

RBC

After about 5 cycles (10 days)
gametocytes form

Sexual cycle
Mosquito — primary host

RBC

Anopheles female

13 day cycle

Mosquito takes up RBCs
with gametocytes

MALARIA LIFE CYCLE

1. Cryptozoite in liver
2. Released merozoite
3. Young trophozoite
4. Older trophozoite
5. Immature schizonts
6. Mature schizonts

7. Merozoites
8. Microgametocyte
9. Macrogametocyte
10. Microgamete
11. Macrogamete
12. Zygote

13. Ookinete
14. Oocyst
15. Oocyst with sporozoites
16. Released sporozoites in
    salivary gland

PARASITIC HELMINTHS

Phylum Platyhelminthes (Flatworms):

A. General Characteristics:

1. Bilaterally symmetrical with leaf-shaped or ribbon-like bodies.

2. Mostly monecious (both sexes in one animal).

3. No body cavity.

4. Digestive tract incomplete or entirely lacking.

5. The parasitic forms are known as tapeworms or flukes.

B. Class Trematoda (Flukes):

1. Characteristics:

a. Unsegmented oval or leaf-shaped bodies with an anterior and ventral sucker.

b. Digestive tract with a single opening that functions as a mouth and an anus.

c. Flukes exist as either ectoparasites or endoparasites.

d. Almost all in the adult stage are parasites of vertebrates (animals and backbones).

2. *Clonorchis sinensis:*

a. This parasite is very important, being responsible for heavy infections in oriental areas where raw fish is a common article of food.

b. Morphology: This fluke is a flat, transparent, flabby , spatulate organism 10–25 mm long and 3.5 mm wide.

1. Digestive Tract: contains mouth, pharynx, esophagus, and digestive caeca.

2. Ventral Sucker: attachment to surface.

3. Testes: produce sperm.

4. Uterus: contains the fertilized eggs.

5. Vitelline or Yolk Glands: produce yolk.

C. Life Cycle:

1. Adults live in the bile ducts of the liver of the host.

2. Eggs pass in the feces and reach water where they remain viable for up to six months. Fluke eggs have either a characteristic spine of an operculum, a cap or lid which is forced open by the hatching embryo.

3. The eggs must be ingested by a certain kind of snail and then hatches in its gut.

4. A ciliated larva called a MIRACIDIUM emerges from the egg, bores into the snail tissue and gives rise to a new generation of SPOROCYSTS.

5. Each sporocyst give rise to a generation of REDIA which give rise to a generation of CERCARIA.

6. The cercaria burrow out of the snail to swim freely in the water.

7. The cercaria must find a fish (carp) within 24–48 hours in which they encyst and become METACERCARIA.

8. The metacercaria remains dormant until the raw fish is ingested by a human.

9. Excystation occurs in the host's duodenum and young fluke migrates to the liver via the bile duct.

D. Clonorchiasis:

1. A fluke disease of the bile ducts.

2. Symptoms:

   a. Results from local irritation of the bile ducts by the flukes from toxemia.

   b. Loss of appetite.

   c. Diarrhea.

   d. Cirrhosis.

   e. Enlargement and tenderness of the liver.

   f. Progressive ascites and edema.

3. Source of Infection:

   a. Human are infected by eating fresh water fish containing encysted larvae.

   b. During digestion, larvae are freed from cysts, and migrate via the bile duct to the liver. Eggs deposited in the bile ducts are passed into the feces of the host.

4. Mode of Infection: Fecal/Oral Route.

5. Diagnosis: Finding the characteristic fluke eggs in the feces.

6. Prevention:

   a. Thorough cooking of all freshwater fish.

   b. Sanitary disposal of infected person's feces.

E. *Schistosoma mansoni*:

1. This organism is a blood fluke. It lives in the veins of the colon of its host. It is found in fresh waters of Africa, South America and the Caribbean.

2. Morphology:

   a. The male worm is 8–16 mm long and 0.5 mm wide. It has a cylindrical appearance due to the fact that the sides of the flat body are folded over to form a ventral grove.

   b. In this grove, projecting freely at each end, but enclosed in the middle, is the longer and more slender female.

   c. Both the male and female worms have oral and ventral suckers. The digestive tract lacks a pharynx.

   d. The eggs are non-operculate and have a lateral spine.

F. Life Cycle:

1. Eggs are laid by the adults in the veins of the colon and are forced through the wall of the gut into the lumen using its characteristic spine.

2. The eggs pass in the feces and hatch immediately in water to give rise to a MIRACIDIUM.

3. The miracidium must penetrate the tissue of a suitable host snail and gives rise to a generation of SPOROCYSTS.

4. Sporocysts produce CERCARIA which swim in water and infect the host by penetrating the skin.

5. The young fluke is carried by the blood to the colon where it matures.

G. Schistosomiasis:

1. A blood fluke infection with the adult worms living in the veins of the host.

2. Symptoms:

   a. Intestinal symptoms.

   b. Portal hypertension and liver involvement.

3. Source and Mode of Infection:
    a. Infection is acquired from water containing larval forms (cercaria) which have developed in snails.
    b. Larvae penetrate human skin, enter the blood and migrate to the liver, where they mature.
    c. Adults migrate to the veins of the gut.
4. Diagnosis:
    a. Presence of eggs in the feces.
5. Prevention:
    a. Disposal of feces and urine so that eggs will not reach bodies of fresh water containing the host snail.
    b. Reduction of snail habitats.

H. Class cestoda:
    1. Characteristics:
        a. No mouth or no digestive tract.
        b. Body ribbon-shaped and composed of 3 regions:
            1. Scolex: May have hooks and or suckers for attaching to the intestinal wall of the host.
            2. Neck: Located just behind the scolex. Products new segments or proglottids.
            3. Proglottids: Sexual reproduction units that are budded off from the neck. The newer proglottids are closest to the to the neck. The older ones are pushed posteriorly. Each proglottid produces first male and then female sex cells.
    2. Tapeworm Diseases:
        a. Symptoms:
            1. When the eggs are swallowed by a human, they hatch into infective larvae that can migrate via the blood to all parts of the body.
            2. The frequency of their occurrence as determined by autopsies is given in the following order: brain, eye, muscle, heart, liver, and lungs.
            3. The gravity of the infection depends on the organ involved and the number of larvae present.
            4. The disease is chronic and may cause serious disability with a relatively high mortality rate.
    3. Source and Mode of Infection: Ingestion of raw or inadequately cooked pork containing the larvae.
    4. Diagnosis: Microscopic examination of feces for eggs of *Taenia solium* or identification of the proglottids.
    5. Prevention:
        a. Thoroughly cook pork.
        b. Adequate inspection of meat at the packing plants.
        c. Prevention of soil and water contamination with human feces in rural areas.
        d. Avoid use of sewage effluents for the irrigation of pastures.

Phylum Nematoda (Roundworms)

A. General Characteristics:
    1. Body thread-like and cylindrical.
    2. Unsegmented.
    3. Sexes are separate in different animals.
    4. A complete digestive tract.

B. *Enterobius vermicularis* (Pinworm): This worm is worldwide in its distribution. In the U.S. it is the most common human worm parasite.

  1. Life Cycle:

     a. The adult lives in the human caecum and colon. When the female is "ripe" she migrates to the anus to the perianal skin, where she lays her eggs and dies. The activities of the female create intense itching.

     b. The eggs are infective when laid. Eggs lodged on the fingers or under the fingernails are carried to the mouth and swallowed. Infection may also result from contaminated bed linen and from inhalation of the eggs floating in the air.

  2. Lab Work:

     a. Obtain a prepared slide of the adult female pinworm (slide #18). Note the long, pointed post-anal TAIL.

     b. Obtain a prepared slide of *Enterobius* ova (slide #19). Note the characteristic shape of the ova. They are transparent and hard to find, but they are there.

  3. Enterobiasis (Pinworm Infection):

     a. Symptoms: Include anal itching and irritability.

     b. Source and Mode of Infection:

        1. Direct transfer of infected eggs by hand. The fecal-oral route.

        2. Indirect through clothing, bedding, and food contaminated by eggs.

        3. Dust borne infection by inhalation.

     c. Diagnosis: Applying transparent adhesive tape to the perianal region and examining it under the microscope for eggs. Material is best obtained in the morning before bathing or defecation.

     d. Prevention:

        1. Daily bathing with showers preferred to tubs.

        2. Frequent changing of underclothing, sheets, and night clothes.

        3. Washing of hands after defecation and always before eating or preparing food.

        4. Removal of sources of infection by treatment of cases with proper medication.

C. *Trichinella spiralis*: This parasite has a worldwide distribution. The main reservoir for human infection is the pig. The incidence of infection of the U.S. as determined by postmortem exams is about 17%.

  1. Life Cycle:

     a. Adults live in the small intestine of the host. The female deposits larvae in the tissues of the small intestine.

     b. The larvae are carried by the blood to skeletal muscle where they form cysts.

     c. Two hosts are required to complete the life cycle. Humans get the disease when they ingest raw or improperly cooked pork.

     d. Humans are a dead end in life cycle because human flesh infected with the larvae is not usually eaten.

  2. Trichinosis:

     a. Symptoms:

        1. Sudden swelling of the upper eyelids.

        2. Muscle soreness and pain.

        3. Thirst, chills, profuse sweating, and weakness.

     b. Source and Mode if Infection: Eating raw or insufficiently cooked flesh of animals (pork, beef, bear) containing encysted larvae.

    c. Diagnosis:

       1. Increasing numbers of blood eosinophils.

       2. Biopsy of skeletal muscle not earlier than 10 days after exposure to the infection.

    d. Prevention:

       1. Inspection of meat at the packing plant by public health staff.

       2. Cook all fresh pork and game meat to a temperature of 150 degrees F.

D. *Necator americanus* (Hookworm): This parasite is commonly known as the New World Hookworm or the "American Killer". It is found in the southern U.S., Central and South America, and Africa. There is an Old World Hookworm, but we will not study it.

  1. Life Cycle:

    a. Adults live in the small intestine of humans. The female produces eggs that pass in the feces and mature in soil.

    b. Larvae hatch from the eggs, and develop into filariform larvae that penetrate the skin to infect the new host.

    c. The larvae migrate via the blood to the lungs of the host where they enter the air spaces.

    d. The larvae are then coughed up, swallowed, and pass to the small intestine.

  2. Hookworm Disease:

    a. Symptoms:

       1. Chronic, debilitating disease with a variety of vague symptoms that vary according to the degree of anemia (due to the blood loss caused by the adult worms).

       2. Children with long-term, heavy infections may be retarded in mental and physical development.

       3. Dermatitis causes itching at the site of infection.

    b. Source and Mode of Infection:

       1. Eggs in feces are deposited on the ground.

       2. Larvae hatch from the eggs and burrow into the skin of the foot.

    c. Diagnosis: Examination of the feces to locate eggs (fecal smear).

    d. Prevention:

       1. Proper sanitation for human feces.

       2. Wear shoes to protect the feet from infection.

LIST OF MEDIA

All of the media listed below are available commercially. Specific instructions concerning the preparation of each medium is included on the label of the bottle or container.

Blood Agar in TSA
Desoxycholate Agar (DLA)
Eosin Methylene Blue Agar (EMB)
Klinger's Iron Agar
Lactose Broth (Single and Double Strength)
Litmus Milk
MacConkey Agar (MAC)
Mannitol Salt Agar (MSA)
m-Endo Broth
MR-VP Broth
Nitrate Broth
Nutrient Agar
Nutrient Broth
Nutrient Gelatin
Phenol Red Broth Base (for Carbohydrate Broths)
Phenylalanine Agar
Phenylethanol Agar (PEA)
Sabouraud Agar
Summons Citrate Agar
SIM medium
Skin Mild Agar
Snyder's Test Agar
Spirit Blue Agar
Starch Agar
Strayne Agar—Developed by Dr. J. Payne & Prepared by Richard Tran at Bergen Community College.
Strayne Broth—Developed by Dr. J. Payne & Prepared by Richard Tran at Bergen Community College
Thioglycollate Agar
Tryptic Soy Agar (TSA)
Tryptic Soy Broth (TSB)
Tryptone Broth
Tryptone Glucose Extract Agar (TGEA)
Urea Broth